SpringerWienNewYork

Ulrich Schubert
Nicola Hüsing
Richard Laine (Eds.)

Materials Syntheses

A Practical Guide

SpringerWienNewYork

Prof. Dr. Ulrich Schubert

Institute of Materials Chemistry, Vienna University of Technology, Vienna, Austria

Prof. Dr. Nicola Hüsing

Inorganic Chemistry I, Ulm University, Ulm, Germany

Prof. Dr. Richard M. Laine

Departments of Chemistry, and Materials Science and Engineering, University of Michigan, Ann Arbor, MI, USA

SpringerWienNewYork is part of
Springer Science+Business Media
springer.at

Typesetting: Camera ready by the editors
Printing: Strauss GmbH, 69509 Mörlenbach, Germany

Printed on acid-free and chlorine-free bleached paper
SPIN: 12123858

Library of Congress Control Number: 2008926905

ISBN 978-3-211-75124-4 SpringerWienNewYork

Contents

Preface

Everywhere one hears complaints that the area of materials synthesis suffers from unclarities, irreproducibility, a lack in detail as well as a lack in standards. The need to remedy this deficiency, which is characteristic of a fast emerging scientific domain, is the main motivation for this book. With the strong and fast development of the world of materials chemistry over the last decades, the need and timeliness for *Materials Syntheses* is clear and urgent. *Materials Syntheses* has the ambition to set standards for documenting materials syntheses.

Materials syntheses are generally more complex than syntheses of inorganic or organic compounds, and the specific characterization methods play a more important role. From the materials point of view, a compound, say TiO_2, can appear as a single crystal, as an amorphous monolith, as a thin film, as nanoparticles etc., and each of these forms requires a completely different preparative route and may have different materials properties, such as surface area, etc. Thus, protocols for materials syntheses and characterizations need to be more diverse than common inorganic or organic synthesis procedures.

A broad variety of different materials classes are represented in this book, ranging from organic polymers to carbonaceous and ceramic materials, from gels to porous and layered materials and from powders and nanoparticles to films. This broad coverage also extends to the preparation methods. Among others, intercalation and flux methods, sol-gel processing, templating methods for porous materials, sonochemistry or spray pyrolysis are represented in this volume. Selection of the contributions was based on techniques that are widely available in materials science laboratories to allow using this book, for example, in materials chemistry laboratory courses at universities.

The *Introduction* of each contribution includes a concise and critical summary of important uses and applications of the described material, as well as key issues for the given procedure. The *Procedures* section provides detailed and unambiguous laboratory directions for the synthesis and application of the specific material. This includes descriptions of the hardware used in as much detail as necessary to allow reproduction of the synthesis with related but not necessarily identical pieces of equipment. Specific characteristics (advantages and disadvantages) of the given procedure, as well as the scope of applicability are also discussed. For example, what synthesis parameters can be changed without changing the general characteristics and the general outcome of the method, and the materials properties? What material properties are influenced to what degree by the synthesis parameters? What are the most common pitfalls in the synthesis? Finally, methods that unequivocally identify the material, characterize its properties and allow for checking as to whether the synthesis was successful are given.

We hope to attract a broad readership, reflecting the diversity of materials science. In addition, all levels will benefit from the book: graduates, post-graduates senior researchers, educators, technicians and scientists working in industry.

The preparation of a book with a new concept does not happen without the input of many people. The Editors wish to express their special thanks to Prof. David Avnir at the Hebrew University of Jerusalem, who represents several others, for many brainstorming sessions. We also thank the contributors to this volume for their patience and willingness to participate in a new project that, if successful, could be continued in additional volumes.

U. Schubert, N. Hüsing and R. M. Laine

List of Contributors

Lidia Armelao
Istituto di Scienze e Tecnologie Molecolari del CNR and INSTM, Dipartimento di Chimica, Università di Padova, Via Marzolo, 1, 35131 Padova, Italy

Davide Barreca
Istituto di Scienze e Tecnologie Molecolari del CNR and INSTM, Dipartimento di Chimica, Università di Padova, Via Marzolo, 1, 35131 Padova, Italy

Carolina Belver
Departamento de Química Inorgánica. Universidad de Salamanca. Plaza de la Merced, S/N. 37008 Salamanca, Spain

Wolfgang H. Binder
Institute of Chemistry, Macromolecular Chemistry, Martin-Luther University Halle-Wittenberg, TGZ-III / Heinrich Damerowstr. 4, 06120 Halle, Germany

Carmen Blanco
Department of Chemical Engineering and Inorganic Chemistry, University of Cantabria, Avda. de los Castros, S/N, ES-39005 Santander, Spain

Timothy J. Boyle
Sandia National Laboratories, Advanced Materials Laboratory, 1001 University Boulevard, SE, Albuquerque, NM 87106, USA

Scott D. Bunge
Department of Chemistry, Kent State University, Kent, OH 44242, USA

Dominik Brühwiler
Institute of Inorganic Chemistry, University of Zürich, Winterthurerstr. 190, 8057 Zürich, Switzerland

Gion Calzaferri
Department of Chemistry and Biochemistry, University of Berne, Freiestr. 3, 3012 Bern, Switzerland

Carmen Canevali
INSTM, Dipartimento di Scienza dei Materiali, Università di Milano-Bicocca, via Cozzi, 53, 20125 Milano, Italy

Kathleen A. Carrado
Chemistry Division, Argonne National Laboratory, 9700 South Cass Av., Argonne, IL 60137, USA

Hao Ming Chen
Department of Chemistry, National Taiwan University, Taipei 106, Taiwan

Norberto Chiodini
INFM – INSTM, Dipartimento di Scienza dei Materiali, Università di Milano-Bicocca, via Cozzi, 53, 20125 Milano, Italy

Ramesh Chitrakar
National Institute of Advanced Industrial Science and Technology, 2217-14 Hayashi-cho, Takamatsu, 761-0395, Japan

Antonio Currao
Department of Chemistry and Biochemistry, University of Bern, Freiestr. 3, 3012 Bern, Switzerland

Nada M. Dimitrijevic
Argonne National Laboratory, 9700 South Cass Av., Argonne, IL 60439, USA

Le-Quyenh Dieu
Institute of Inorganic Chemistry, University of Zürich, Winterthurerstr. 190, 8057 Zürich, Switzerland

Delphine Fargier
Departments of Chemistry, and Materials Science and Engineering, and the Macromolecular Science and Engineering Center, University of Michigan, Ann Arbor, MI 48109-2136, USA

Qi Feng
Department of Advanced Materials Science, Faculty of Engineering, Kagawa University, 2217-20 Hayashi-cho, Takamatsu, 761-0396, Japan

Mário José Ferreira Calvete
Institut für Organische Chemie II, Universität Eberhard-Karls Tübingen, Auf der Morgenstelle 18, 72076 Tübingen, Germany

Andreas O. Gabriel
Merck KGaA, Frankfurter Str. 250, 64293 Darmstadt, Germany

Jasmin Geserick
Inorganic Chemistry I, Ulm University, Albert-Einstein-Allee 11, 89081 Ulm, Germany

Antonio Gil
Departamento de Química Aplicada, Universidad Pública de Navarra, 31006 Pamplona, Spain

Fernando González
Department of Chemical Engineering and Inorganic Chemistry, University of Cantabria, Avda. de los Castros, S/N, 39005 Santander, Spain

Antonella Glisenti
Università degli Studi di Padova, Dipartimento di Scienze Chimiche, Via Marzolo 1, 35131 Padova, Italy

Silvia Gross
ISTM-CNR, Dipartimento di Scienze Chimiche, Università di Padova, via Marzolo, 1, 35131 Padova, Italy

Michael Hanack
Institut für Organische Chemie II, Universität Eberhard-Karls Tübingen, Auf der Morgenstelle 18, 72076 Tübingen, Germany

Dieter Holzinger
Institute of Materials Chemistry, Vienna University of Technology, Getreidemarkt 9, 1060 Wien, Austria

Nicola Hüsing
Inorganic Chemistry I, Ulm University, Albert-Einstein-Allee 11, 89081 Ulm, Germany

Sorin Ivanovici
Institute of Materials Chemistry, Vienna University of Technology, Getreidemarkt 9, A-1060 Wien, Austria

Fares Khairallah
Dubai BioTechnology and Research Park (DuBiotech), P.O.Box 73000, Dubai, United Arab Emirates

Guido Kickelbick
Institute of Materials Chemistry, Vienna University of Technology, Getreidemarkt 9, 1060 Wien, Austria

Seung-Gyoo Kim
Departments of Chemistry, and Materials Science and Engineering, and the Macromolecular Science and Engineering Center, University of Michigan, Ann Arbor, MI 48109-2136, USA

Joachim Köhler
Inorganic Chemistry I, Ulm University, Albert-Einstein-Allee 11, 89081 Ulm, Germany

Richard M. Laine
Departments of Chemistry, and Materials Science and Engineering, and the Macromolecular Science and Engineering Center, University of Michigan, Ann Arbor, MI 48109-2136, USA

Claudia Leiggener
Department of Chemistry and Biochemistry, University of Bern, Freiestrasse 3, 3012 Bern, Switzerland

Christian Lembacher
Institute of Materials Chemistry, Vienna University of Technology, Getreidemarkt 9, 1060 Wien, Austria

Ru-Shi Liu
Department of Chemistry, National Taiwan University, Taipei 106, Taiwan

Zong-huai Liu
Key Laboratory of Applied Surface and Colloid Chemistry, Ministry of Education, School of Chemistry and Materials Science, Shaanxi Normal University, Xi'an, 710062, China

Mariachiara Mattoni
INSTM, Dipartimento di Scienza dei Materiali, Università di Milano-Bicocca, via Cozzi, 53, 20125 Milano, Italy

Franca Morazzoni
INSTM, Dipartimento di Scienza dei Materiali, Università di Milano-Bicocca, via Cozzi, 53, 20125 Milano, Italy

Saifun Nahar-Borchert
Clariant GmbH, Am Unisyspark 1, 65843 Sulzbach, Germany

Kenta Ooi
National Institute of Advanced Industrial Science and Technology, 2217-14 Hayashi-cho, Takamatsu, 761-0395, Japan

Carmen Pesquera
Department of Chemical Engineering and Inorganic Chemistry, University of Cantabria, Avda. de los Castros, S/N, 39005 Santander, Spain

Tijana Rajh
Argonne National Laboratory, 9700 South Cass Av., Argonne, IL 60439, USA

Ralf Riedel
Fachbereich Material- und Geowissenschaften, Fachgebiet Disperse Feststoffe, Technische Universität Darmstadt, Petersenstrasse 23, 64287 Darmstadt, Germany

Arantzazu Zabala Ruiz
Department of Chemistry and Biochemistry, University of Berne, Freiestr. 3, 3012 Bern, Switzerland

Annabeth Ryder
School of Chemical Sciences, University of Illinois at Urbana-Champaign, 600 S Mathews Av., Urbana, IL 61801, USA

Giselle Sandí
Chemistry Division, Argonne National Laboratory, 9700 South Cass Av., Argonne, IL 60137, USA

Zoran V. Saponjic
Argonne National Laboratory, 9700 South Cass Av., Argonne, IL 60439, USA

Irina Sapurina
Institute of Macromolecular Compounds, Russian Academy of Sciences, V.O. Bolshoi pr. 31, St. Petersburg 199004, Russia

Ulrich Schubert
Institute of Materials Chemistry, Vienna University of Technology, Getreidemarkt 9, 1060 Wien, Austria

Roberto Scotti
INSTM, Dipartimento di Scienza dei Materiali, Università di Milano-Bicocca, via Cozzi, 53, 20125 Milano, Italy

Sara E. Skrabalak
School of Chemical Sciences, University of Illinois at Urbana-Champaign, 600 S. Mathews Av., Urbana, IL 61801, USA

Jeroen Spooren
Dipartimento di Chimica Industriale ed Ingegneria dei Materiali, Universita' di Messina, Salita Sperone 31, 98166 Messina, Italy

Jaroslav Stejskal
Institute of Macromolecular Chemistry, Academy of Sciences of the Czech Republic, Heyrovsky Sq. 2, 162 06 Prague 6, Czech Republic

Santy Sulaiman
Departments of Chemistry, and Materials Science and Engineering, and the Macromolecular Science and Engineering Center, University of Michigan, Ann Arbor, MI 48109-2136, USA

Kenneth Suslick
School of Chemical Sciences, University of Illinois at Urbana-Champaign, 600 S Mathews Av., Urbana, IL 61801, USA

Weiping Tang
Research Institute for Solvothermal Technology, 2217-43 Hayashi-cho, Takamatsu, 761-0301, Japan

Gregor Trimmel
Institute of Materials Chemistry, Vienna University of Technology, Getreidemarkt 9, 1060 Wien, Austria

Miguel Ángel Vicente
Departamento de Química Inorgánica, Universidad de Salamanca, Plaza de la Merced, S/N, 37008 Salamanca, Spain

Richard I. Walton
Department of Chemistry, University of Warwick, Coventry, CV4 7AL, UK

Randall E. Winans
Chemistry Division, Argonne National Laboratory, 9700 South Cass Av., Argonne, IL 60137, USA

Xiaojing Yang
College of Chemistry, P.O. Box S-46, Beijing Normal University, Beijing, 100875, China

Ronald Zirbs
Institute of Chemistry, Macromolecular Chemistry, Martin-Luther University Halle-Wittenberg, TGZ-III / Heinrich Damerowstr. 4, 06120 Halle, Germany

Controlling Size and Morphology of Zeolite L

A. Z. Ruiz, D. Brühwiler, L.-Q. Dieu and G. Calzaferri

Abstract The synthesis of zeolite L crystals of high purity and well-defined morphology is described. Four procedures are detailed, leading to (a) large elongated cylindrical crystals, (b) medium-sized disc-shaped crystals, (c) nano-sized crystals, and (d) medium-sized cylindrical-shaped crystals.

Classification

form:	crystalline powder
function:	molecular sieve, host material for supramolecular organization of organic molecules
preparation:	hydrothermal synthesis
composition:	$M_9[Al_9Si_{27}O_{72}]\cdot n\ H_2O$ ($M = K^+$ or Na^+)

Introduction

Zeolites are crystalline aluminosilicates featuring defined channels and cavities.[1] The ability to accommodate various organic and inorganic species while being transparent in the UV-Vis-NIR makes zeolites ideal host materials for supramolecular organization.[2] In many cases, the confinement of molecules in zeolites and the catalytic activity of surface adsorption sites lead to interesting photochemical phenomena not observed in solution.[3]

Zeolite L possesses one-dimensional channels arranged in a hexagonal pattern. The free diameter of the main channels varies from 7.1 Å (narrowest part) to 12.6 Å (widest part). The morphology of the crystals can be approximated by a cylinder, with the channel entrances located at the base planes. A crystal with a diameter of 550 nm typically consists of about 80'000 parallel channels. High-resolution electron microscopy has been used to image the surface structures of zeolite L and to advance the understanding of growth processes and defects.[4] Materials obtained by the inclusion of organic dye molecules into the channels of zeolite L feature a variety of intriguing properties, ranging from increased dye stability to photonic antenna functions and optical anisotropy.[5]

In most procedures for zeolite synthesis, the gel composition is given as a ratio of oxides. In order to correctly calculate the required amounts of starting material from this ratio, one has to take into account that hydroxides can be considered oxides plus water (KOH = ½ K_2O and ½ H_2O, for example). The purity of the starting materials should also be considered, bearing in mind that the water content of the hydroxides can be quite significant. Our procedure to calculate the amounts of starting material for a given molar composition

$$a\,K_2O–b\,Na_2O–c\,Al_2O_3–d\,SiO_2–e\,H_2O$$

is as follows. The required amounts of KOH, NaOH, $Al(OH)_3$, SiO_2, and H_2O are:

$$\mathrm{Weight}_{\mathrm{KOH}} = \frac{\mathrm{Mol}_{\mathrm{KOH}} \cdot \mathrm{MW}_{\mathrm{KOH}}}{\mathrm{P}_{\mathrm{KOH}}} \quad \text{with} \quad \mathrm{Mol}_{\mathrm{KOH}} = 2a$$

$$\mathrm{Weight}_{\mathrm{NaOH}} = \frac{\mathrm{Mol}_{\mathrm{NaOH}} \cdot \mathrm{MW}_{\mathrm{NaOH}}}{\mathrm{P}_{\mathrm{NaOH}}} \quad \text{with} \quad \mathrm{Mol}_{\mathrm{NaOH}} = 2b$$

$$\mathrm{Weight}_{\mathrm{Al(OH)_3}} = \frac{\mathrm{Mol}_{\mathrm{Al(OH)_3}} \cdot \mathrm{MW}_{\mathrm{Al(OH)_3}}}{\mathrm{P}_{\mathrm{Al(OH)_3}}} \quad \text{with} \quad \mathrm{Mol}_{\mathrm{Al(OH)_3}} = 2c$$

$$\mathrm{Weight}_{\mathrm{SiO_2}} = \mathrm{Mol}_{\mathrm{SiO_2}} \cdot \mathrm{MW}_{\mathrm{SiO_2}} \quad \text{with} \quad \mathrm{Mol}_{\mathrm{SiO_2}} = d$$

$$\mathrm{Weight}_{\mathrm{TotH_2O}} = \mathrm{Mol}_{\mathrm{H_2O}} \cdot \mathrm{MW}_{\mathrm{H_2O}} \quad \text{with} \quad \mathrm{Mol}_{\mathrm{TotH_2O}} = e$$

MW designates the respective molecular weights, whereas P represents the purities. This means that 98 % pure KOH has a $P_{KOH} = 0.98$. The total amount of water is distributed among the silica suspension and the aluminate solution as follows:

$$\mathrm{Weight}_{\mathrm{H_2O\ SiO_2 susp.}} = \frac{\mathrm{Weight}_{\mathrm{SiO_2}}\left(1-\mathrm{P}_{\mathrm{SiO_2}}\right)}{\mathrm{P}_{\mathrm{SiO_2}}}$$

$$\begin{aligned}\mathrm{Weight}_{\mathrm{H_2O\ Al_2O_3 sol.}} = \mathrm{Weight}_{\mathrm{TotH_2O}} - \Big[&\mathrm{Weight}_{\mathrm{H_2O\ SiO_2 susp.}} + \mathrm{Weight}_{\mathrm{KOH}}\left(1-\mathrm{P}_{\mathrm{KOH}}\right) \\ &+ \mathrm{Weight}_{\mathrm{NaOH}}\left(1-\mathrm{P}_{\mathrm{NaOH}}\right) + \mathrm{Weight}_{\mathrm{Al(OH)_3}}\left(1-\mathrm{P}_{\mathrm{Al(OH)_3}}\right) \\ &+ \frac{1}{2}\cdot \mathrm{Mol}_{\mathrm{KOH}} \cdot \mathrm{MW}_{\mathrm{H_2O}} + \frac{1}{2}\cdot \mathrm{Mol}_{\mathrm{NaOH}} \cdot \mathrm{MW}_{\mathrm{H_2O}} \\ &+ \frac{3}{2}\cdot \mathrm{Mol}_{\mathrm{Al(OH)_3}} \cdot \mathrm{MW}_{\mathrm{H_2O}}\Big]\end{aligned}$$

P_{SiO_2} specifically refers to the SiO_2 content of the silica suspension.

Materials

- Ludox HS-40 (40 wt.% SiO_2), purchased from Aldrich, used as received.

- Aerosil OX-50 (silica powder), purchased from Degussa, used as received.
- Aerodisp W 1226 (26 wt.% SiO_2), purchased from Degussa, used as received.
- Aluminum hydroxide, purchased from Fluka, purity >99 %, used as received.
- Aluminum powder, purchased from Fluka, purity >99 %, used as received.
- Potassium hydroxide pellets, purchased from Fluka, purity ≥86 %, used as received.
- Sodium hydroxide pellets, purchased from Merck, purity >99 %, used as received.
- Doubly distilled water (used throughout the synthesis).
- Pressure-tight poly(tetrafluoroethylene) (PTFE) vessel (see ref. 6 for an example).

Safety and Disposal

Safety and handling instructions for the chemicals are found in the corresponding materials safety data sheets (MSDS). The pressure-tight PTFE vessel should be designed to well withstand the vapor pressure of water at the employed temperatures (at least 175 °C which corresponds to approximately 890 kPa).

Procedures

A. Large Zeolite L Crystals (Fig. 1)

An amount of 3.11 g of potassium hydroxide is added to 22.00 g of doubly distilled water and stirred at 0 °C (on ice) for 5 min. Next, 0.58 g of metallic aluminum powder is added under nitrogen flow, and the mixture is stirred at 0 °C for 15 min. After letting the solution warm to room temperature, stirring is continued for 1.5 h under nitrogen flow. The resulting solution is filtered to remove $Fe(OH)_3$, which is due to Fe as an impurity in aluminum, until a clear solution is obtained. This solution is added to 14.34 g of Ludox under vigorous stirring, the latter having been stirred between 5 and 10 min beforehand. After 3 to 6 min of further stirring, the opaque gel is transferred to the PTFE vessel for crystallization at 175 °C for 72 h under static conditions. The composition of the gel is 2.24 K_2O–1.00 Al_2O_3–8.98 SiO_2–164.40 H_2O.

After crystallization, the pressure vessel is cooled in ice for 1 h before opening. The product is centrifuged (4000 rpm, 8 min) and washed with boiling doubly distilled water until the pH of the supernatant becomes neutral. The crystals are dried for approximately 16 h at 80 °C in air, yielding about 5 g of material. Subsequent ion exchange is performed by suspending the material in 70 ml of doubly distilled water and adding 4.0 g of potassium nitrate. After stirring this suspension for 5 h at about 50 °C, the zeolite is centrifuged (4000 rpm, 8 min) and washed until the pH of the supernatant becomes neutral. Finally, the crystals are dried in air for 16 h at 80 °C.

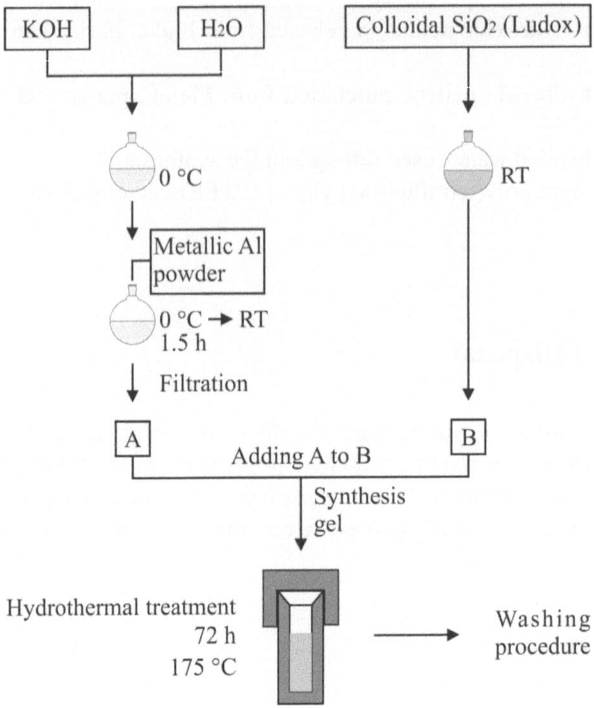

Fig. 1. Scheme of the synthesis procedure for large zeolite L crystals.

Characterization

The following experiment was performed to conveniently check the success of the synthesis. When zeolite L is added to an aqueous solution of thionine, aggregates of the dye immediately form on the external zeolite surface. Upon boiling the sample for about 1 min, a sudden color change from violet to blue is observed. This effect is due to the insertion of the dye molecules into the zeolite L channels where they can exist as monomers only. The blue color remains after cooling to room temperature. If the test is negative (no color change), the additional characterization methods are unnecessary.[7]

The products were analyzed by X-ray powder diffraction (using a Guinier camera de Wolff Mk.IV, CuK_{α} radiation, ENRAF-NONIUS and also a Stoe STADIP powder diffractometer in transmission, $CuK_{\alpha 1}$ radiation, focusing Ge(III) monochromator) for phase identification (Fig. 2). The patterns were compared to a standard pattern of commercial zeolite L (Union Carbide or UOP). Reference XRD patterns are also available from ref. 8. The morphology of the crystals was examined by means of scanning electron microscopy (JEOL JSM 840 and Hitachi S-3000N). A homogeneous distribution of cylindrical crystals with hexagonal cross-section and smooth surfaces is obtained. The average length of the crystals is 6 μm with an average diameter of 2 μm.

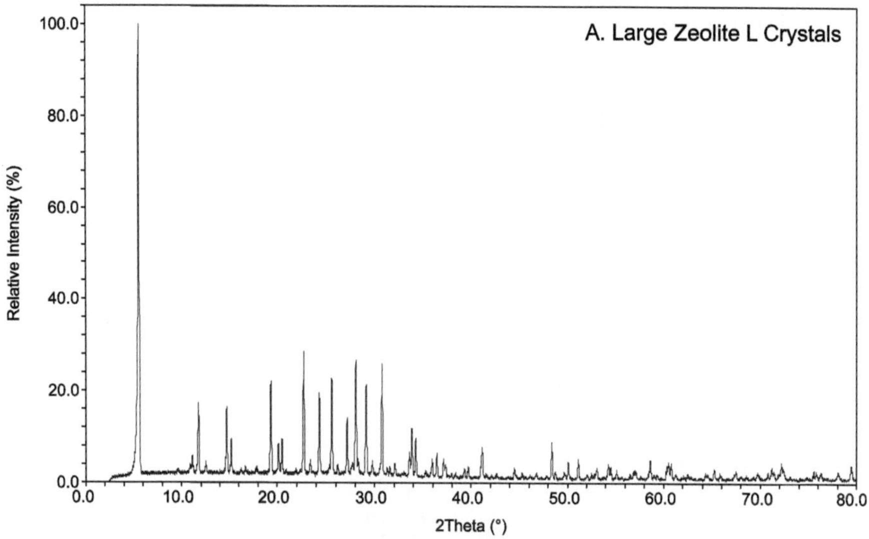

Fig. 2. X-ray powder diffractogram for large zeolite L crystals measured in transmission.

Comments

The reason for ion exchange is to have only potassium ions as exchangeable cations. The final composition is obtained after the ion exchange.

B. Medium-sized Disc-shaped Zeolite L Crystals (Fig. 3)

An amount of 2.76 g of potassium hydroxide, 1.74 g of sodium hydroxide, and 0.62 g of aluminum hydroxide are added to 17.40 g of doubly distilled water and refluxed for 3 h in an oil bath at 120 °C resulting in a clear solution. After letting this solution cool to room temperature, it is added under stirring to 17.67 g of Ludox, the latter having been stirred between 5 and 10 min beforehand. After 3 to 6 min of further stirring, the opaque gel is transferred to the PTFE vessel for crystallization at 160 °C for 48 h under dynamic conditions (rotation at 40 rpm). The composition of the gel is 5.40 K_2O–5.50 Na_2O–1.00 Al_2O_3–30.00 SiO_2–416.08 H_2O. For the application of dynamic crystallization conditions, an oven equipped with a device enabling rotation of the PTFE vessels at various speeds is used (see ref. 6 for details).

After crystallization, the pressure vessel is cooled in ice for 1 h before opening. The product is centrifuged (4000 rpm, 8 min) and washed with boiling doubly distilled water until the pH of the supernatant becomes neutral. The crystals are dried for approximately 16 h at 80 °C in air, yielding about 2 g of material. Ion exchange is performed as outlined in procedure A.

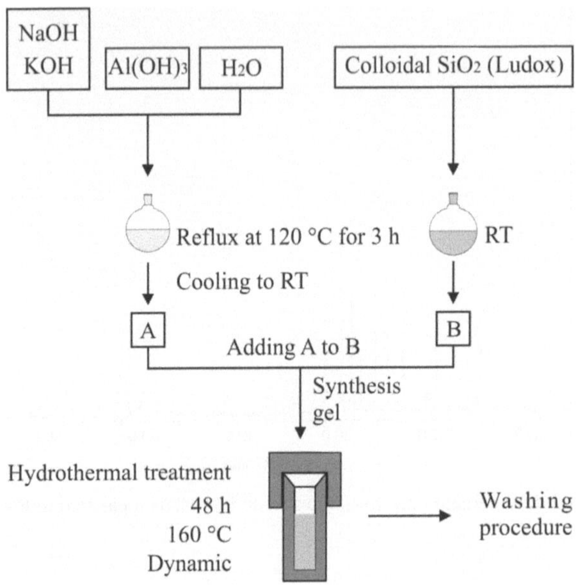

Fig. 3. Scheme of the synthesis procedure for medium-sized disc-shaped zeolite L crystals.

Characterization

Characterization was performed as described for Synthesis A. The X-ray powder diffractogram is shown in Fig. 4. A homogeneous distribution of disc-shaped zeolite L crystals with an average length of 0.35 μm and an average diameter of 1 μm is obtained.

Comments

Partial addition of sodium hydroxide yields disc-shaped crystals of higher quality by reducing intergrowth. Aging the final gel in the closed PTFE vessel for 15 h at room temperature before crystallization leads to crystals with an average length of 0.065 μm and an average diameter of 0.4 μm.

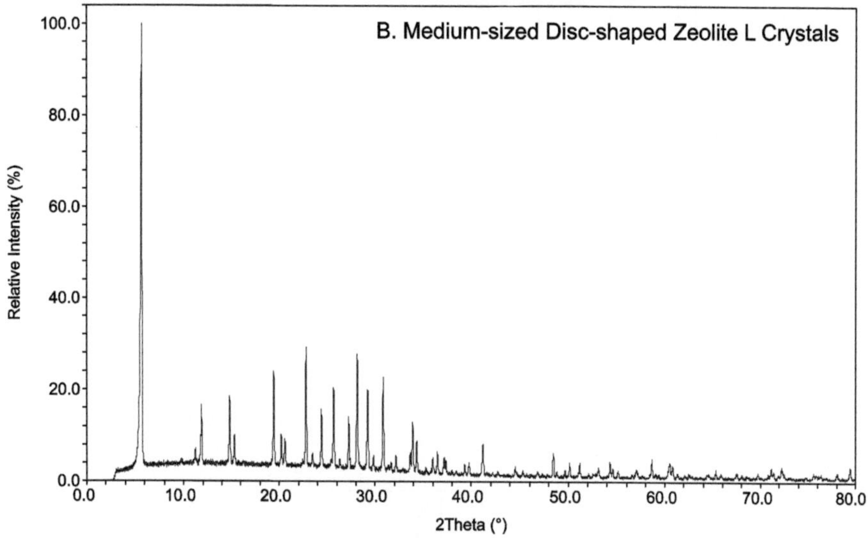

Fig. 4. X-ray powder diffractogram for medium-sized disc-shaped zeolite L crystals measured in transmission.

C. Nano-sized Zeolite L Crystals (Fig. 5)

An amount of 4.84 g of potassium hydroxide and 1.56 g of aluminum hydroxide is added to 20.00 g of doubly distilled water and refluxed for 15 h in an oil bath at a temperature of 115 °C resulting in a clear solution. A silica suspension is prepared separately as follows: 28.04 g of doubly distilled water are added to 12.02 g of silica powder and suspended for 15 min at 18000 rpm (Ultra Turrax mixer, IKA T18 Basic). This suspension is left between 30 min and 1 h, and mixed for 10 min at 18000 rpm before use.

An amount of 7.23 g of potassium hydroxide and 21.68 g of doubly distilled water is added to the silica suspension and refluxed for 15 h in an oil bath at 115 °C. After letting the potassium aluminate solution and the potassium silica suspension cool to room temperature, the potassium aluminate solution is added to the potassium silica suspension under vigorous stirring. After stirring for 3 to 6 min, the opaque gel is transferred to the PTFE vessel for crystallization at 170 °C for 6 h under dynamic conditions (rotation at 16 rpm). The composition of the gel is 9.34 K_2O–1.00 Al_2O_3–20.20 SiO_2–412.84 H_2O.

After crystallization, the pressure vessel is cooled in ice for 1 h before opening. The product is centrifuged (5000 rpm, 40 min) and washed with boiling doubly distilled water until the pH of the supernatant becomes neutral. The crystals are dried for approximately 16 h at 80 °C in air yielding about 1.5 g of material. Ion exchange is performed as outlined in procedure A.

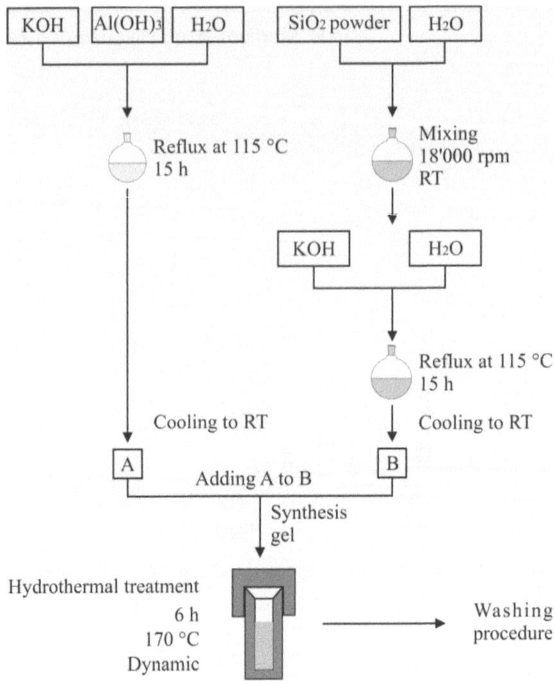

Fig. 5. Scheme of the synthesis procedure for nano-sized zeolite L crystals.

Characterization

The products were analyzed by XRD (Fig. 6) and transmission electron microscopy (Hitachi H-600-2 and Philips XL30 ESEM-FEG). Zeolite L crystals with dimensions in the order of 30 nm are obtained, featuring a tendency to agglomerate into larger clusters of 80–100 nm. The XRD pattern shows the line broadening expected for such small crystallites.

D. Medium-sized Cylindrical-shaped Zeolite L Crystals (Fig. 7)

An amount of 3.18 g potassium hydroxide, 1.60 g of sodium hydroxide, and 2.21 g of aluminum hydroxide is added to 9.40 g of doubly distilled water and refluxed for 3 h in an oil bath at 120 °C resulting in a clear solution. A silica suspension is prepared separately as follows: 35.07 g of colloidal silica (Aerodisp W1226, Degussa, 26 wt.% of SiO_2) and 6.39 g of doubly distilled water are mixed and kept in an ultrasonic bath for about 10 min. After letting the potassium sodium aluminate solution cool to room temperature, it is added to the colloidal silica suspension under vigorous stirring. After further stirring for 3 min, the opaque gel is transferred to the PTFE vessel for crystallization at 160 °C for 144 h under static conditions. The composition of the gel is 1.73 K_2O—1.41 Na_2O—1.00 Al_2O_3 —10.81 SiO_2—173.00 H_2O.

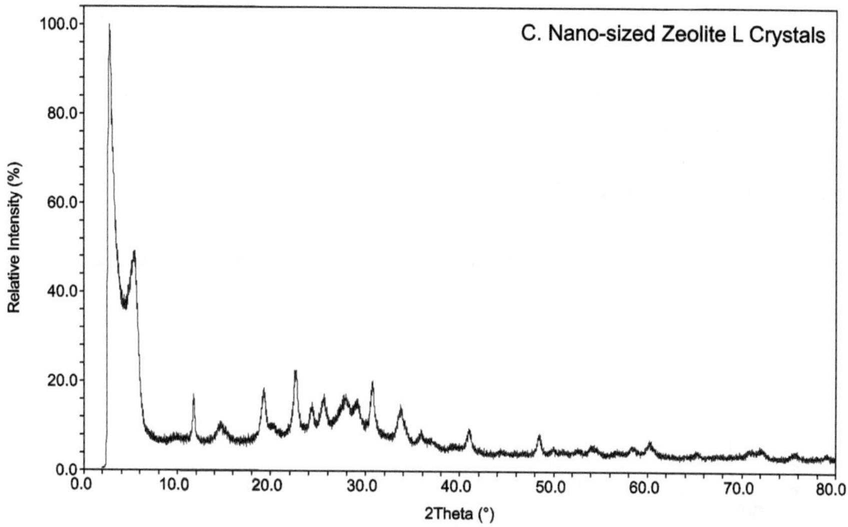

Fig. 6. X-ray powder diffractogram for nano-sized zeolite L crystals measured in transmission.

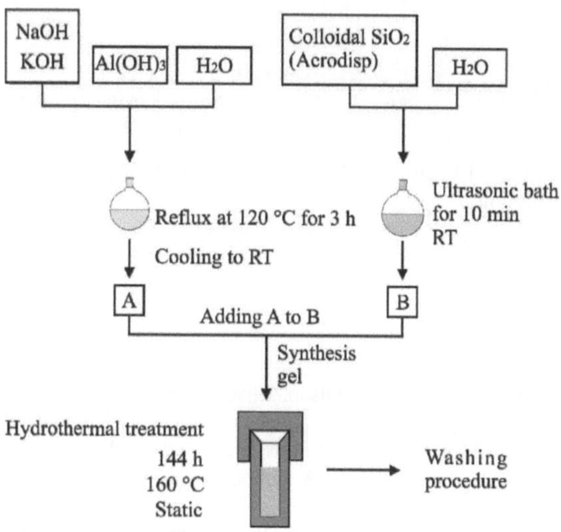

Fig. 7. Scheme of the synthesis procedure for medium-sized cylindrical-shaped zeolite L crystals.

After crystallization, the pressure vessel is cooled in ice for 1 h before opening. The product is centrifuged (4000 rpm, 8 min) and washed with boiling doubly distilled water until the pH of the supernatant becomes neutral. The crystals are dried for approximately 16 h at 80 °C in air yielding about 6 g of material. Ion exchange is performed as outlined in procedure A.

Characterization

Characterization was performed as described for Synthesis A. The X-ray diffractogram is shown in Fig. 8. A homogeneous distribution of medium-sized zeolite L crystals with smooth surfaces and an average length of 0.9 μm and an average diameter of 0.7 μm is obtained.

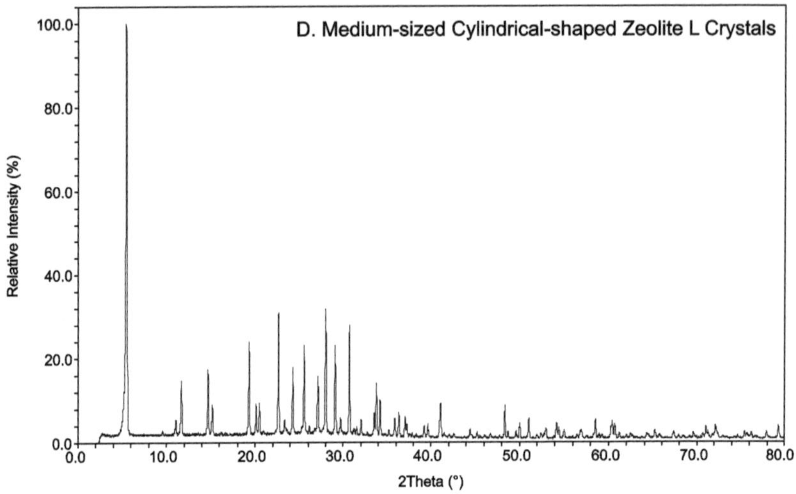

Fig. 8. X-ray powder diffractogram for medium-sized cylindrical-shaped zeolite L crystals measured in transmission.

References

[1] C. Baerlocher, W. M. Meier, D. H. Olson, Atlas of Zeolite Framework Types, Elsevier, 2001 (see also: http://www.iza-structure.org /databases).
[2] D. Brühwiler, G. Calzaferri, Microporous Mesoporous Mater. **72**, 1 (2004). G. Schulz-Ekloff, D. Wöhrle, B. van Duffel, R. A. Schoonheydt, Microporous Mesoporous Mater. **51**, 91 (2002).
[3] S. Hashimoto, J. Photochem. Photobiol. C: Photochem. Rev. **4**, 19 (2003).
[4] T. Ohsuna, B. Slater, F. Gao, J. Yu, Y. Sakamoto, G. Zhu, O. Terasaki, D. E. W. Vaughan, S. Qiu, C. R. A. Catlow, Chem. Eur. J. **10**, 5031 (2004). O. Terasaki, T. Ohsuna, Top. Catal. **24**, 13 (2003).
[5] O. Bossart, L. De Cola, S. Welter, G. Calzaferri, Chem. Eur. J. **10**, 5771 (2004). T. Ban, D. Brühwiler, G. Calzaferri, J. Phys. Chem. B **108**, 16348 (2004). G. Calzaferri, S. Huber, H. Maas, C. Minkowski, Angew. Chem. Int. Ed. **42**, 3732 (2003). A. Zabala Ruiz, H. Li, G. Calzaferri, Angew. Chem. Int. Ed. **45**, 5282 (2006).
[6] A. Zabala Ruiz, D. Brühwiler, T. Ban, G. Calzaferri, Monatsh. Chem. **136**, 77 (2005).

[7] G. Calzaferri, D. Brühwiler, S. Megelski, M. Pfenniger, M. Pauchard, B. Hennessy, H. Maas, A. Devaux, U. Graf, Solid State Sci. **2**, 421 (2000).
[8] M. M. J. Treacy, J. B. Higgins, R. von Ballmoos, Collection of Simulated XRD Powder Diffraction Patterns of Zeolites, Elsevier, 2001 (see also: http://www.iza-structure.org/databases).

Zeolite A and ZK-4

C. Leiggener, A. Currao and G. Calzaferri

Abstract The synthesis of zeolite A and ZK-4 crystals of high purity and well-defined morphology is described. Three procedures are detailed, leading to cubic crystals of zeolite A with chamfered edges (average size 3–5 μm), cubic crystals of ZK-4 with sharp edges (average size 1–2 μm), and nano-sized cubic crystals of zeolite A with slightly rounded edges (size ≤ 1 μm).

Classification

form:	crystalline powder
function:	molecular sieve, host material for supramolecular organization of quantum dots
preparation:	hydrothermal synthesis
composition:	$Na_{12}[(AlO_2)_{12}(SiO_2)_{12}]\cdot 27H_2O$ (zeolite A), $Na_9[(AlO_2)_9(SiO_2)_{15}]\cdot nH_2O$ (ZK-4)

Introduction

Classical zeolites are crystalline aluminosilicates, consisting of an anionic framework and charge-compensating cations.[1,2] The primary building units of the framework are SiO4 and AlO4 tetrahedra. The framework is build from corner-sharing TO4 tetrahedra (T = Si, Al) leading to microporous materials featuring defined channels and cavities. The presence of aluminum results in a negatively charged framework, which is compensated by protons or cations inside the cavities. Additional water molecules can also be present in the cavities under ambient conditions. Zeolites are used in a broad range of applications. Due to their ion exchange capability they can act as water softeners or be used for the removal of pollutants, and their well defined cavities allow size-selective reactions, for example in catalysis.[3] Being transparent in the UV/Vis/NIR makes zeolites ideal host materials for supramolecular organization of different kinds of molecules, clusters, and metal complexes.[4]

The procedures described here allow the convenient synthesis of zeolite A and ZK-4. Both zeolites have the same structure with a 3-dimensional channel system and a channel opening around 4.1 Å (Fig. 1). The main difference between zeolite A and ZK-4 is the chemical composition, i.e. a different Si/Al ratio (see below), and therefore they have a different number of charge compensating cations. As materials with the same structure but with a different chemical composition they belong to the same framework type (framework code LTA, see Ref. 1 for more details). Two kinds of structural subunits are formed.[1] The smaller consists of 24 T-atoms and is commonly denoted as β-cage, sodalite cage, or pseudo-unit cell. Eight β-cages are linked by four-membered rings giving rise to a larger cavity called α-cage with 48 T-atoms and a diameter around 11.4 Å. Consequently, the framework can also be build from face-sharing α-cages connected by eight-membered rings. The resulting 3-dimensional channel system in zeolite A and ZK-4 turned out to be convenient for hosting small ions and semiconductor clusters.[5,6]

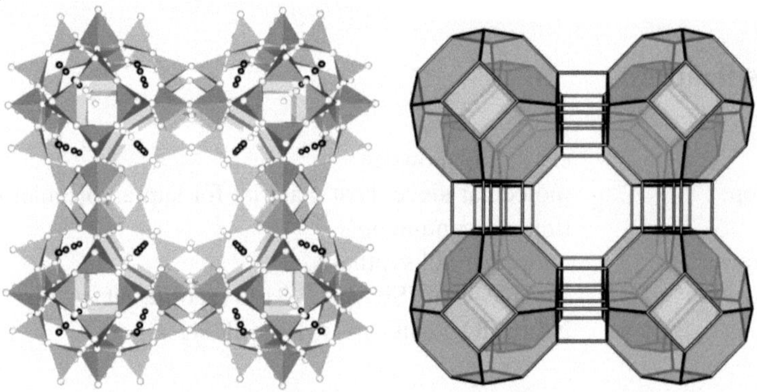

Fig. 1. Left: View of the structure of zeolite A and ZK-4. Oxygen: white spheres; T-positions (Si, Al): center of tetrahedra; cation (Na): black spheres. Right: Framework of zeolite A and ZK-4 (framework code LTA). In framework representations, the bridging oxygen atoms are usually omitted and a straight line is drawn between T-atoms. Gray polyhedra: β-cage.

Convenient synthesis procedures for pure zeolite A and ZK-4 by sol-gel methods were developed to obtain chloride-free microcrystals of high chemical quality, very good crystallinity and size homogeneity.[7] Modification of the composition of the starting gel, especially the Si/Al ratio and the Na^+ content, results in different morphologies of the crystals or alteration in the size distribution (Fig. 2). A lower Na^+ content in the gel generally leads to smaller crystals, while an excess of Si leads to ZK-4. The synthesis procedures presented here typically yield crystals with the chemical composition $Na_{12}[(AlO_2)_{12}(SiO_2)_{12}]\cdot 27H_2O$ for zeolite A and $Na_9[(AlO_2)_9(SiO_2)_{15}]\cdot nH_2O$ for ZK-4, respectively. The number n of water molecules per pseudo-unit cell depends on the cation. The water content in ZK-4 was not determined. The Na^+ can be replaced by other monovalent or divalent cations by means of ion exchange.[6] For zeolite A the Si/Al ratio is around 1. For ZK-4 the ratios is around 1.7. This increased silicon content in ZK-4 results in a small contraction of the unit cell parameters.[1,2] In procedure A, tetraethoxysilane ($Si(OEt)_4$), aluminium, and sodium hydroxide (NaOH) were used. In procedure B and C a certain amount of NaOH was replaced by tetramethylammoniumhydoxide (TMAOH) in order to reduce the Na^+ content. Very fine SiO_2 powder was used as silicon source. The TMA-ions were removed by calcination after the synthesis. Afterwards, full sodium loading can be obtained by ion exchange of protons with Na^+ from sodium nitrate ($NaNO_3$) solution.

Fig. 2. SEM images of the products obtained by synthesis procedure A (cubic crystals of zeolite A with chamfered edges, average size 3–5 µm), procedure B (cubic crystals of ZK-4 with sharp edges, average size 1–2 µm), and procedure C (nano-sized cubic crystals of zeolite A with slightly rounded edges, size ≤ 1 µm).

Materials

- Tetraethoxysilane, $Si(OEt)_4$, purchased from Aldrich, purity >99 %, used as received.
- Diisopropylamine purchased from Merck, used as received.
- Aluminium wire (1 mm) purchased from Balzers, purity 99.999 %.
- Sodium hydroxide (NaOH) pellets purchased from Merck, purity >99 %, used as received.
- Tetramethylammoniumhydroxide (TMAOH) purchased from Aldrich, used as received.
- Doubly distilled water (used throughout the synthesis).
- Ethanol absolute (99.8 %).
- Sodium nitrate ($NaNO_3$) purchased from Merck, purity p.a., used as received.
- Pressure-tight polytetrafluoroethylene (PTFE) vessel (see Ref. 8 for details).
- For the application of dynamic crystallization conditions in procedure C, a drying oven equipped with a device was used enabling rotation of the PTFE vessels at various speeds (see Ref. 8 for details).

Safety and Disposal

Safety and handling instructions for the chemicals are found in the corresponding materials safety data sheets (MSDS). The pressure-tight PTFE vessel should be designed to well withstand the vapor pressure of water at the employed temperatures (at least 175°C, which corresponds to approximately 890 kPa or 9 bar).

Procedures

In all procedures, two solutions, solution 1 and 2, are prepared separately in Teflon flasks. Solution 1 contains the Si precursor and solution 2 the Al precursor. The two solutions are combined forming a gel. Heating the gel for several hours leads to zeolite crystal formation. Size, morphology, and final composition of the crystals depend mainly on the composition and ageing of the starting gels. The silicon source in procedure A is tetraethoxysilane, while in procedures B and C SiO_2 is used (see Fig. 3 and 4).

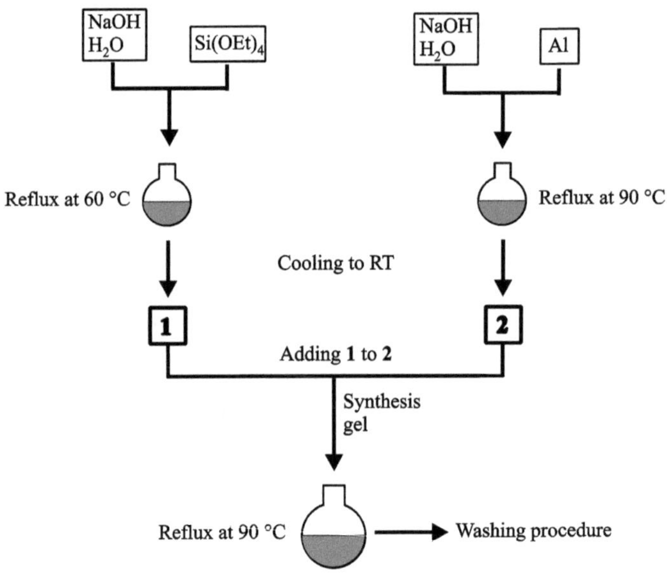

Fig. 3. Scheme of the synthesis procedure for zeolite A with NaOH (procedure A).

Very fine SiO_2 powder is obtained as follows: 72.86 g of diisopropylamine is diluted in 75 ml doubly distilled water and stirred for 1 h until the solution is clear. An amount of 150 g of tetraethoxysilane is added under vigorous stirring, and the mixture is stirred for 48 h. The product is centrifuged (4500 rpm, 20 min) and the white sediment suspended in 400 ml of ethanol. The suspension is stirred for 1 h and then filtrated (glass frit, pore size 4). The product is washed twice with ethanol and dried in vacuum at 100°C for 2 – 3 h. Thermogravimetric analysis gives a loss of 20 % at 800°C.

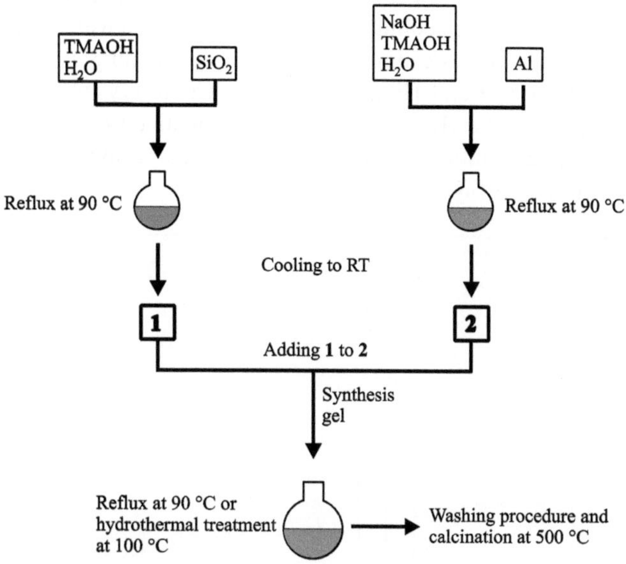

Fig. 4. Scheme of the synthesis procedure for ZK-4 and nano-sized zeolite A with NaOH/TMAOH (procedures B and C).

A. Zeolite A Crystals (Fig. 3)

Solution 1: 5.93 g of NaOH is dissolved in 150 ml of doubly distilled water and 7.721 g of $Si(OEt)_4$ is added. The mixture is refluxed under nitrogen atmosphere at 60 °C for 3 h under stirring. *Solution 2:* 5.93 g of NaOH is dissolved in 150 ml doubly distilled water and 2 g of Al-wire is added. The mixture is refluxed under nitrogen atmosphere at 90 °C for 3 h under stirring.

The two clear solutions are cooled to room temperature before solution 1 is added to solution 2 under stirring. The gel is stirred for 15 min at room temperature and then for 16 h under reflux at 90°C. The mixture is transferred into a beaker for sedimentation of the product. The supernatant liquid is carefully removed, and the zeolite crystals are washed three times with 250 ml of boiling doubly distilled water each time. Separation is done by centrifugation (4500 rpm, 15 min). The product is washed with ethanol, filtrated (glass frit, pore size 4) and dried at 80°C in an oven.

Characterization

The product was analyzed by X-ray powder diffraction, XRD, for phase identification (STOE STADI P, transmission mode, $CuK_{\alpha1}$ radiation) (Fig. 5). The pattern was compared to a standard pattern of commercial zeolite A (Union Carbide). A reference XRD pattern for zeolite A is also available from Ref. 9. The morphology of the crystals (Fig. 2) was examined by means of scanning electron microscopy, SEM, (JOEL JSM 840 and Hitachi S-3000N). Cubic crystals of zeolite A with chamfered edges and nearly no intergrowth are obtained (average size 3–5 μm). The composition determined by means of energy dispersive X-ray spectroscopy, EDX, is $Na_{12}[(AlO_2)_{12}(SiO_2)_{12}]$.

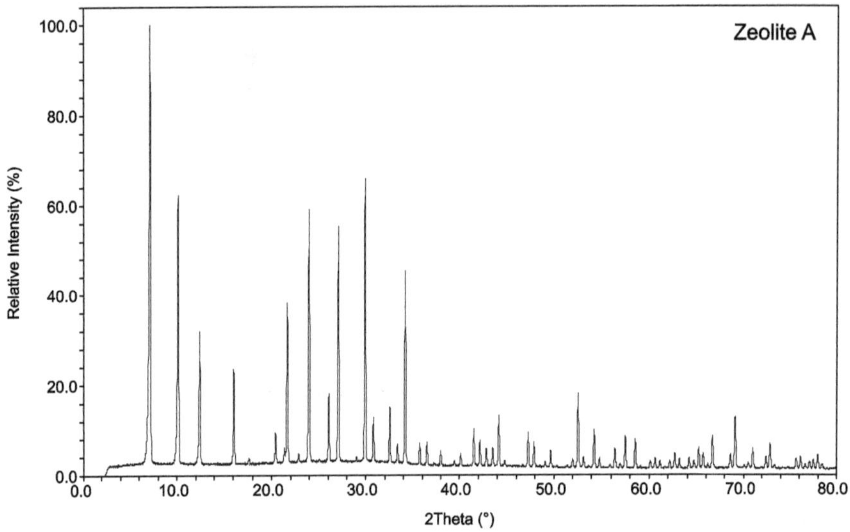

Fig. 5. X-ray powder diffraction diagram of zeolite A synthesized according to the procedures above.

B. ZK-4 Crystals (Fig. 4)

Solution 1: 4.34 g of SiO_2 and 25.74 g of TMAOH are dissolved in 57 ml of doubly distilled water. The mixture is refluxed under nitrogen atmosphere at 90 °C for 2 h under stirring. *Solution 2:* 0.962 g of Al-wire, 2 g of NaOH, and 6.435 g of TMAOH are dissolved in 50 ml doubly distilled water. The mixture is refluxed under nitrogen atmosphere at 90°C for 3 h under stirring.

The two clear solutions are cooled to room temperature before solution 1 is added to solution 2 under stirring. The gel is stirred for 15 min at room temperature and then for 24 h under reflux at 90°C. The product is separated by centrifugation (4500 rpm, 15 min). The white sediment is washed by suspending it three times in boiling doubly distilled water (250 ml) and each time centrifuged. After washing with ethanol the product is first dried at 80°C in an oven. The TMA ions are removed by calcination at 500°C for 16 h in air. Afterward, full sodium loading is obtained by ion exchange of protons with Na^+ by suspending the zeolites three times for 15 min in 0.1 M $NaNO_3$ solution.

Characterization

Characterization was performed as described for zeolite A. Cubic crystals of ZK-4 with sharp edges (average size 1–2 μm). Composition determined by EDX: $Na_9[(AlO_2)_9(SiO_2)_{15}]$ (Fig. 2). The X-ray diffractogram is shown in Fig. 6.

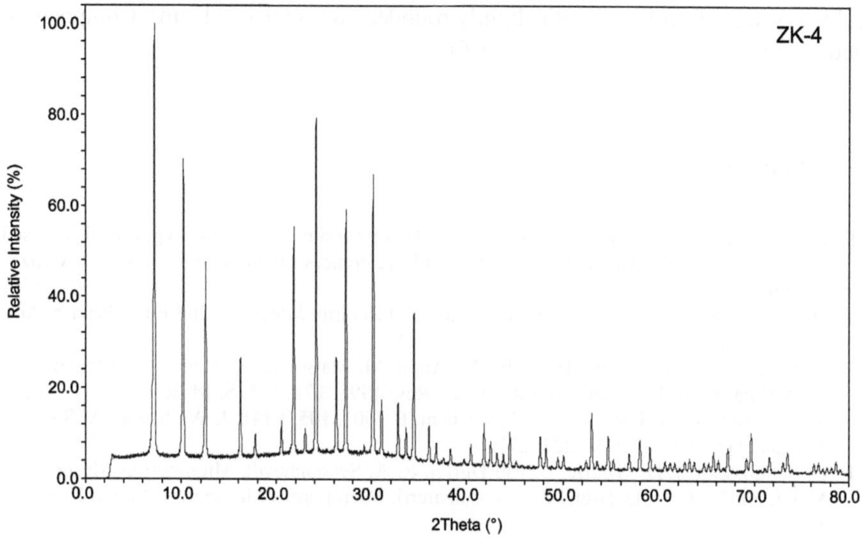

Fig. 6. X-ray powder diffraction diagram of zeolite ZK-4 synthesized according to the procedures above.

C. Nano-sized Zeolite A Crystals (Fig. 4)

Solution 1: 0.6225 g of SiO_2 and 4.47 g of TMAOH are dissolved in 57 ml of doubly distilled water. The mixture is refluxed under nitrogen atmosphere at 90 °C for 2 h under stirring. *Solution 2:* 0.5 g of Al-wire, 0.6 g of NaOH, and 5.328 g of TMAOH are dissolved in 50 ml of doubly distilled water. The mixture is refluxed under nitrogen atmosphere at 90°C for 4 h under stirring.

The two clear solutions are cooled to room temperature before solution 1 is added to solution 2 under stirring. The gel is stirred for 2 d at room temperature. For crystallization, the gel is filled into a PTFE vessel and placed in an oven equipped with a device enabling rotation(rotation at 40 rpm) for 3 d at 100°C. The pressure vessel is then cooled in ice for 1 h before opening. The product is separated by centrifugation (5000 rpm, 30 min) and washed three times with boiling doubly distilled water, centrifuging each time. After washing with ethanol the product is first dried at 80°C in an oven. The TMA ions are removed by calcination at 500°C for 16 h in air. Afterward, full sodium loading is obtained by ion exchange of protons with Na^+ by suspending the zeolites three times for 15 min in 0.1 M $NaNO_3$ solution.

Characterization

Characterization was performed as described above (Figs. 2 and 5). Nano-sized cubic crystals of zeolite A with slightly rounded edges (size ≤ 1 μm). Composition determined by EDX: $Na_{12}[(AlO_2)_{12}(SiO_2)_{12}]$

References

[1] C. Baerlocher, W. M. Meier, D. H. Olson, Atlas of Zeolite Framework Types, Fifth Revised Edition, Elsevier, Amsterdam, 2001, and references therein (see also: www.iza-structure.org).

[2] H. Robson, K. P. Lillerud, Verified Syntheses of Zeolitic Materials, 2nd Ed., Elsevier, Amsterdam, 2001 (see also: www.iza-synthesis.org).

[3] S. T. King, J. Catal. 1996, **161**, 530. M. Anpo, M. Matsuoka, K. Hanou, H. Mishima, H. Yamashita, H. H. Patterson, Coord. Chem., Rev. 1998, **171**, 175. S. M. Kanan, C. P. Tripp, R. N. Austin, H. H. Patterson, J. Phys. Chem. B 2001, **105**, 9441. J. Weitkamp, A. Raichle, Y. Traa, Appl. Cat. A 2001, **222**, 277.

[4] G. Schulz-Ekloff, D. Wöhrle, B. van Duffel, R. A. Schoonheydt, Microporous Mesoporous Mater. 2002, **51**, 91. D. Brühwiler, G. Calzaferri, Microporous Mesoporous Mater. 2004, **72**, 1.

[5] W. Sachtler, Acc. Chem. Res., 1993, **26**, 383. M. Wark, G. Schulz-Ekloff, N. I. Jaeger, Bulg. Chem. Comm. 1998, **30**, 129. A. A. Demkov, O. F. Sankey, J. Phys.: Cond. Matter 2001, **13**, 10433; d) K. Kuge, G. Calzaferri, Microporous Mesoporous Mater. 2003, **66**, 15.

[6] D. Brühwiler, R. Seifert, G. Calzaferri, J. Phys. Chem. B 1999, **103**, 6397. D. Brühwiler, C. Leiggener, S. Glaus, G. Calzaferri, J. Phys. Chem. B 2002, **106**, 3770. C. Leiggener, D. Brühwiler, G. Calzaferri, J. Mater. Chem. 2003, **13**, 1969. C. Leiggener, G. Calzaferri, ChemPhysChem 2004, **5**, 1593. M. Meyer, C. Leiggener, G. Calzaferri, ChemPhysChem 2005, **6**, 1071. M. Meyer, A. Currao, G. Calzaferri, ChemPhysChem 2005, **6**, 2167. C. Leiggener, G. Calzaferri, Chem. Eur. J. 2005, **11**, 7191.

[7] P. Lainé, R. Seifert, R. Giovanoli, G. Calzaferri, New J. Chem. 1997, **21**, 453. R. Seifert, R. Rytz, G. Calzaferri, J. Phys. Chem. A 2000, **104**, 7473.

[8] A. Zabala Ruiz, D. Brühwiler, T. Ban, G. Calzaferri, Monatsh. Chem. 2005, **136**, 77.

[9] M. M. J. Treacy, J. B. Higgins, Collection of Simulated XRD Powder Patterns for Zeolites, 4th Ed., Elsevier, Amsterdam, 2001 (see also: www.iza-structure.org).

Mesostructured Silica Thin Films

J. Köhler, J. Geserick and N. Hüsing

Abstract Mesoporous films composed of amorphous silica with periodically arranged pores are promising materials as catalyst supports, sensors or filtration membranes. Solvent evaporation-induced self-assembly methods based on supramolecular organization of amphiphilic molecules in combination with sol-gel processing of condensable inorganic precursors result in porous thin films with a monomodal pore size distribution. These films are typically deposited by spin coating, casting or dip coating.

Classification

form:	amorphous silica, thin film
function:	porous coating
preparation:	surfactant templating, sol-gel processing
composition:	SiO_2

Introduction

In the early 1990s, researchers discovered that in addition to single molecules such as tetramethylammonium bromide used for the preparation of zeolites, molecular assemblies, as found in liquid crystals, can be used for templating inorganic matrices.[1,2] With this discovery, research in the field of templating and patterning inorganic materials to get perfectly periodic, regularly sized and shaped cavities, channels, and layers in the mesoporous regime (pores with 2–50 nm diameter) has expanded dramatically. This supramolecular templating relies on the ability of amphiphilic molecules to self-assemble into micellar structures that, when concentrated in aqueous solutions, undergo a second stage of self-organization resulting in lyotropic liquid crystal-like mesophases. Molecular inorganic species can cooperatively co-assemble with these structure-directing agents (templates) to eventually condense and form the mesoscopically ordered inorganic backbone of the final material (Fig. 1). The mesostructured nanocomposite is typically either calcined, ozonolyzed or solvent extracted to obtain a porous inorganic material in which the pore dimension relates approximately to the chain length of the hydrophobic tail of the template molecule.

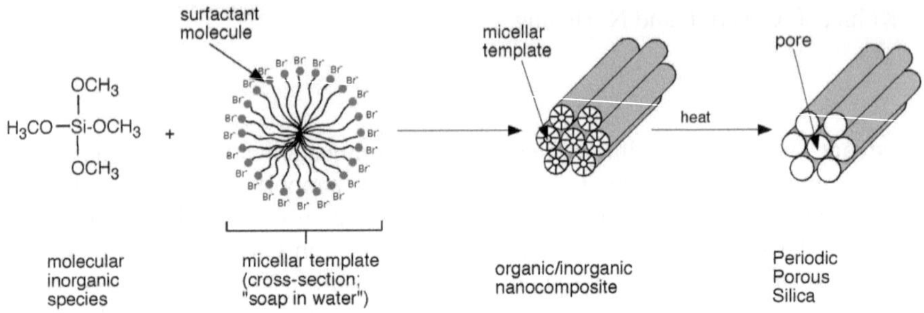

Fig. 1. Schematic presentation of the supramolecular templating

Based on this cooperative self-organization process of inorganic and organic entities not only powders can be formed, but also thin films. Mesostructured thin films can be prepared as free-standing layers or supported by a variety of different substrates, from silica and other inorganic compositions and even as inorganic-organic hybrid materials. Different synthetic approaches can be applied such as growth at interfaces (vapor – liquid, solid – liquid and liquid – liquid), electrodeposition, pulsed-laser deposition techniques or processes based on solvent evaporation techniques. The various synthetic approaches and the formation mechanisms of mesoporous silica films have been reviewed in detail.[3,4]

The procedure described here relies on evaporation-induced self-assembly of solutions containing an inorganic precursor, an organic template molecule that shows the ability to organize in supramolecular arrays, some additives (e.g. acid or base catalysts to start hydrolysis and condensation reactions of the inorganic precursors) and a volatile solvent. A number of different synthesis protocols have been developed by now. The first detailed mechanistic study on dip-coated silica-based samples was performed by Brinker and his coworkers, who termed the process "evaporation-induced self-assembly" (the EISA-process).[5,6] Beginning with a homogeneous solution of ethanol, water, hydrochloric acid, soluble silica source, and surfactant, in a concentration far below the concentration where micelles or other aggregates are formed, preferential evaporation of alcohol during withdrawal of the substrate from the sol concentrates the film in water, silica species and surfactant. Therefore, the surfactant concentration is progressively increasing, resulting in the formation of micelles and upon further evaporation of ethanol in the formation of liquid crystal-like mesophases consisting of silica–surfactant co-assemblies. This process allows the formation of a mesostructured nanocomposite film within a few tens of seconds (Fig. 2).

After template removal, a mesoporous material is obtained in which the pores can be arranged in a hexagonal but also in a cubic fashion. This mechanism was utilized in casting, spin coating and dip coating processes and a variety of films differing in composition, pore size and pore orientation have been synthesized.[7-10]

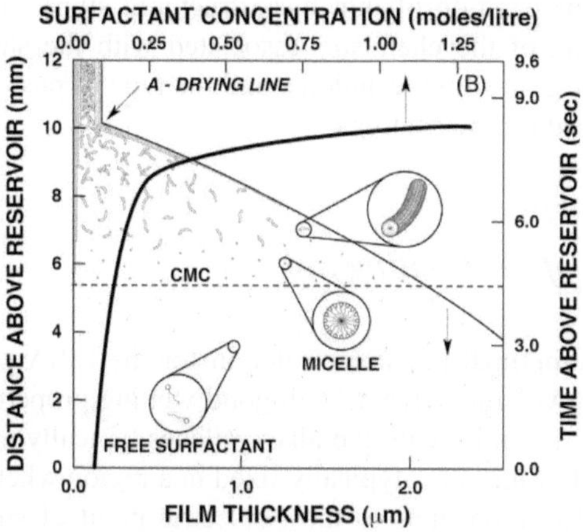

Fig. 2. Schematic representation of the dip coating process.

In the following, one representative synthesis protocol for a mesostructured silica film based on the non-ionic surfactant Brij 56 and tetraethylorthosilicate (TEOS) is presented. Possible modifications and variations of the protocol are mentioned in the text.

Materials

- Silicon wafers as substrates, from Wacker Chemie AG
- Tetraethylorthosilicate (TEOS) from Merck (>98%), used as received.
- Brij 56, (Poly(ethylene glycol)hexadecylether, $C_{16}H_{33}(OCH_2CH_2)_nOH$, (n~10) (Merck), used as received.
- Ethanol (Merck) purity >99,5%, used as received.
- Toluene (Merck, >99,9%)
- Hydrochloric acid (Merck, 32% PA), used as received.

Safety and Disposal

Safety and handling instructions for the chemicals are found in the corresponding materials safety data sheets (MSDS).

Procedures

The reproducible preparation of thin mesostructured silica films requires a profound understanding of the chemistry associated with the starting sol, the processes linked to the deposition technique, and the type of post-treatment (thermal, washing or UV) used for the coatings.

A. Preparation of the Coating Sol

For silica sols, the optimal precursor solution for the EISA process is based on ethanol as a highly volatile solvent, with good wetting properties for hydrophilic substrates and its miscibility with the alkoxysilanes typically applied as inorganic precursors. The pH of the sol is typically fixed in a region where fast hydrolysis is favored over condensation, thus to the isoelectric point of silica (H^+/Si = 0.14-0.003). As an ideal acid, hydrochloric acid has been identified, due to its high volatility upon drying of the film. In addition, the molar ratio of surfactant / inorganic precursor determines the final mesostructure, which can be lamellar, cubic or hexagonal. Another important synthesis parameter is the aging time of the precursor solution, since the degree of condensation of the inorganic species strongly influences the degree of mesostructuring.

Prehydrolyzed solution. The precursor was prepared following a procedure described by Brinker et al. in inert atmosphere (Argon) by mixing 61.00 mL (0.275 mol) of $Si(OEt)_4$ (TEOS) in 61.00 mL ethanol (1.045 mol), 4.87 mL deionized water (0.270 mol) and 0.20 mL of 0.07 M HCl ($1.400 \cdot 10^{-5}$ mol) in a three - necked flask (250 mL).[5] The H^+/Si ratio is therefore set to $5.09 \cdot 10^{-5}$ and the hydrolysis ratio h = 1.03. The solution was heated up to 60 °C for 90 minutes to promote hydrolysis of the alkoxysilane to the corresponding pre-hydrolysed species ($Si(OEt)_{4-x}(OH)_x$), but to minimize the degree of condensation. After cooling to room temperature (still in inert atmosphere), this solution was transferred to a refrigerator at -20°C. The silica sol remains stable for several weeks upon cooling, but condensation reactions continuously lead to an increase in the degree of condensation.

Coating solution. In a typical procedure, 10.00 mL (containing $2.16 \cdot 10^{-2}$ mol Si) of the previously prepared stock solution were dissolved in 20.00 mL of ethanol (0.343 mol). To this solution 0.80 mL ($5.6 \cdot 10^{-5}$ mol) of an aqueous hydrochloric acid solution (0.07 M), thus resulting in a total amount of $5.75 \cdot 10^{-5}$ mol H^+ and 0.80 mL water ($4.44 \cdot 10^{-2}$ mol), were added under stirring. Subsequently, the surfactant – here the non-ionic poly(ethylene glycol)-hexadecylether, Brij 56 – was added to a 10.00 g equivalent of this mixture. This quantity has been varied between 1 and 20 wt%.

B. Film Deposition

Films were deposited by dip or spin coating on silicon wafers. The silicon wafers were cut into 10×20 mm pieces with a diamond pen. Prior to film deposition, the wafer substrates were cleaned by sonification in an ultrasound bath (VWR, Ultrasonic Cleaner) in different solvents, following the protocol listed below:

- sonification in ethanol (15 min)
- sonification in toluene (15 min)
- sonification in ethanol (15 min)
- drying at room temperature

The dip coater was a modified Czochralski apparatus with stageless tunable withdrawal speed in a range between 0–750 mm·min^{-1}. For film deposition the withdrawal speed was set to 240 mm·min^{-1}. Alternatively, films were coated with a Laurell Technologies Corporation Model WS-400B-6NPP/LITE spin coater, by 3000–6000 rpm for 1 min (in Ar atmosphere).

For film deposition via dip coating, the relative humidity (RH) has been identified as a crucial parameter for the formation of the final mesostructure. It has been shown that the mesostructuring can easily be varied by adjusting the relative humidity, which indicates that the quantity of water in the final film depends on the RH. For films prepared in the presence of non-ionic surfactants, the relative humidity is typically set to approx. 35–50%, for ionic surfactants it is set to 0–5% RH. For Brij 56 templated films described here, the RH was set to 40%.

C. Post Treatment

The films were calcined in air at 350°C for 3 h after deposition, using a ramp rate of 1 K·min^{-1}, to promote the prosecution of hydrolysis and condensation reactions to complete the 3D silica network formation and to remove the surfactant.

With the post-treatment step, typically a stabilization of the inorganic network and the formation of porosity by template removal are desired. This is achieved by thermal decomposition of the template phase, since the inorganic mesostructured films show good temperature stability up to 600–1000°C depending on the synthesis conditions. There is a great variety of possible post-treatments that can be applied to the films such as solvent extraction of the organic phase, oxygen plasma treatment, mild temperature treatment, acid or base treatment, or combinations hereof.

Characterization

The silica films prepared by the abovementioned protocol are characterized with respect to their structure by N_2-sorption, X-ray diffraction, and transmission electron microscopy. The coating thickness of 250–400 nm was estimated from SEM images.

X-ray diffraction (XRD). X-ray diffractograms were collected on a diffractometer with a CuK_α source equipped with a high temperature sample chamber in the low angle regime (2θ 1.0–10.0°). The X-ray diffraction patterns can be collected prior to or after the post-treatment (with a high temperature stage even during calcination) and give clear information on the mesostructure. Fig. 3 shows the evolution of the mesostructure during the final heat treatment step of a Brij 56 templated film with 5 wt% of the template in the coating sol.

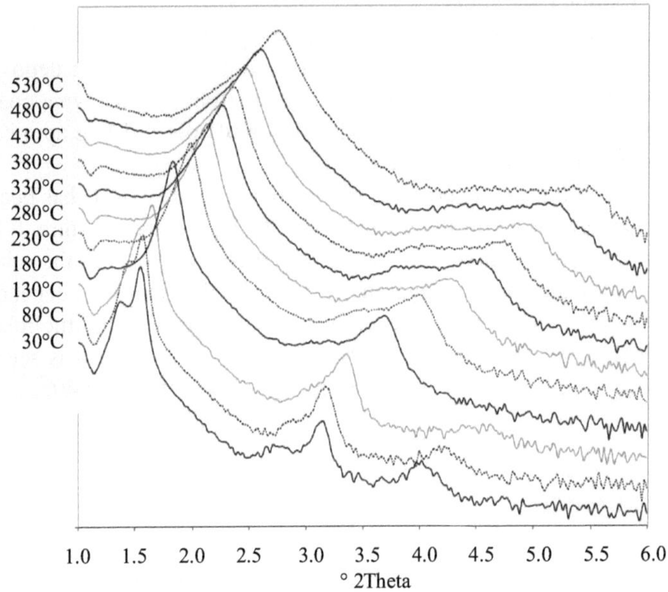

Fig. 3. Temperature-dependent X-ray diffraction patterns of a sample containing 5 wt% Brij 56.

Besides the fact that the film shows mesotexturation, two other aspects can be seen. First, the mesostructure formed after deposition is not homogeneous and can be indexed as two hexagonal phases with different repeating unit distances (at 30 °C: d_{100}= 6.41 nm and 5.69 nm und d_{200}= 3.16 nm and 2.80 nm), but only one phase remains after heating the sample to ~200 °C (at 530 °C: d_{100} = 3.21 nm and d_{200} = 1.61 nm). Second, strong shrinkage of almost 40% of the repeating unit distance is observed during heat treatment, due to the thermally driven network condensation and loss of the organic template. This shrinkage is typically associated with a unidirectional contraction of the meso-domain normal to the surface plane. Fig. 4 shows the X-ray diffraction patterns for samples with different concentrations of the template Brij 56 after calcination. A clear dependence of the degree of mesostructuring and the *d*-spacing of the repeating unit distances on the amount of template is visible.

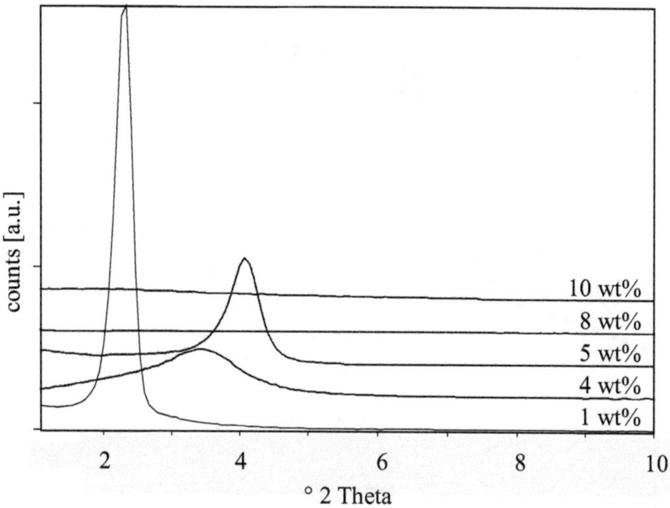

Fig. 4. X-ray diffraction patterns for films templated with different amounts of Brij 56.

N_2-Sorption (BET / BJH). The gas adsorption measurements were performed on a Quantachrome NOVA 4000e and Autosorp MP1. Prior to the measurement, the samples were heated at 350 °C for 3 h under vacuum. The BET surface area was evaluated using adsorption data in a relative pressure range (p/p_0) 0.05–0.2 (S^{BET}). The mesopore size distribution was calculated on the basis of desorption branches using the BJH model (Fig. 5). For N_2-sorption measurements a large amount of sample is needed. Since thin films are difficult to measure, the samples were prepared by casting the solution into a cascade of Petri-dishes and scraping off after drying. The resulting BET surface area varies between 500 and 800 $m^2\,g^{-1}$ for the different silica films. Fig. 5 shows the corresponding isotherms and the insert the pore size distribution calculated from the BJH theory.

The mesostructured silica films show a maximum in the pore diameter distribution at 2.5 nm by templating with Brij 56. Together with the results obtained by XRD, the thickness of the pore-walls can be calculated by $2d_{100}^{(XRD)}/\sqrt{3} - D^{BJH}$ assuming a hexagonal orientation of the mesostructure. Table 1 shows the obtained data for surface area, pore volume, pore wall thickness, etc.

Transmission Electron Microscopy (TEM). TEM measurements were performed on a Philips CM20 after preparing the sample as cross-section. The lower part of the image corresponds to the silicon wafer substrate and the upper part to the glue used to form the sandwich layer. Fig. 6 shows a TEM image of a sample with 5% Brij 56 after dip coating.

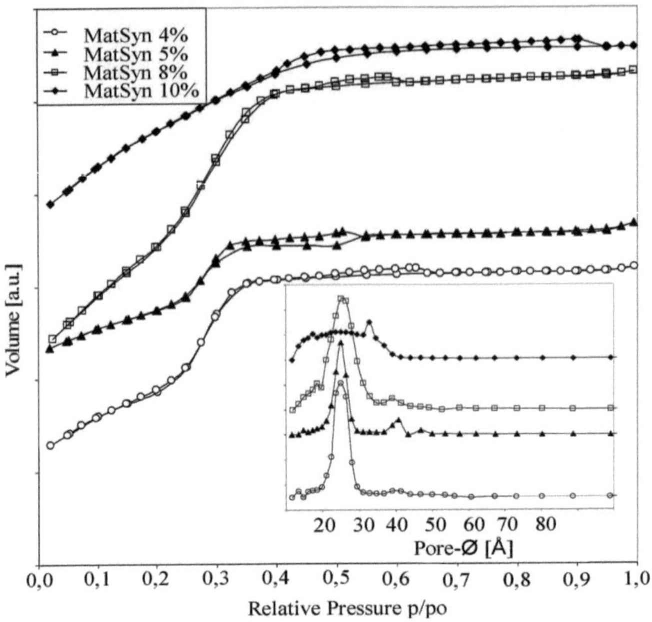

Fig. 5. N_2-sorption isotherms (BJH) for samples with different amounts of the template Brij 56. The insert corresponds to the pore size distribution of these samples.

Table 1. Structural data obtained for samples with different concentrations of the template Brij 56.

Sample	d_{100} /nm	S^{BET} /$m^2\,g^{-1}$	V_{Max}^{N2} /$cm^3\,g^{-1}$	D^{BJH} (w) /nm	t_{wall} /nm
1%	4.09	28	0.4	2.49	1.60
4%	3.21	816	319	2.51	0.70
5%	2.90	530	217	2.51	0.39
8%	-	1122	450	2.52	-
10%	-	789	307	2.52	-

The TEM image verifies the data obtained from X-ray diffraction. The high degree of ordering can be seen in Fig. 6. Since the sample was calcined at 450°C, it can be assumed that the image shows a side view along the c-axis of a hexagonal mesophase. The layer thickness after calcination is given with 130 nm and the repeating unit distance with 2.3 nm.

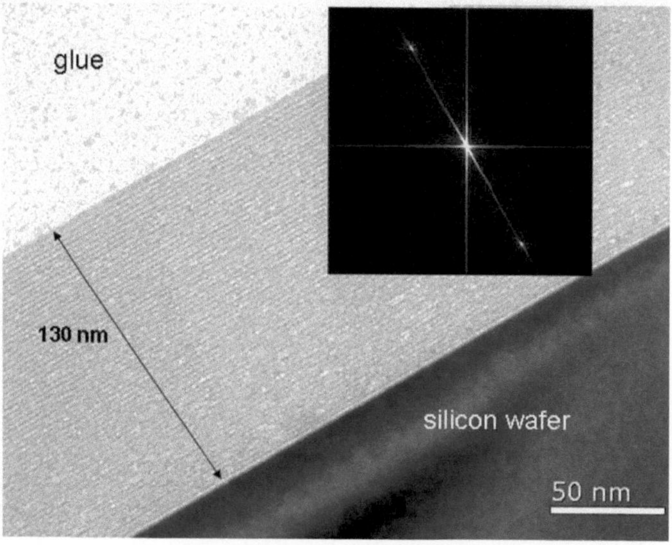

Fig. 6. Cross-sectional TEM image containing 5 wt% Brij 56.

References

[1] J. S. Beck, J. C. Vartuli, W. J. Roth, M. E. Leonowicz, C. T. Kresge, K. D. Schmitt, C. T.-W. Chu, D. H. Olson, E. W. Sheppard, S. B. McCullen, J. B. Higgins, J. L. Schlenker, J. Am. Chem. Soc. 1992, **114**, 10834; C. T. Kresge, M. E. Leonowicz, W. J. Roth, J. C. Vartuli, J. S. Beck, Nature 1992, **359**, 710.

[2] T. Yanagisawa, T. Shimizu, K. Kuroda, C. Kato, Bull. Chem. Soc. Jpn. 1990, **63**, 988.

[3] K. J. Edler, S. J. Roser, Int. Rev. Phys. Chem. 2001, **20**, 387.

[4] L. Nicole, C. Boissière, D. Grosso, A. Quach, C. Sanchez, J. Mater. Chem. 2005, **15**, 5398.

[5] Y. Lu, R. Ganguli, C. A. Drewien, M. T. Anderson, C. J. Brinker, W. Gong, Y. Guo, H. Soyez, B. Dunn, M. H. Huang, J. I. Zink, Nature 1997, **389**, 364.

[6] C. J. Brinker, Y. Lu, A. Sellinger, H. Fan, Adv. Mater. 1999, **11**, 579.

[7] D. Zhao, P. Yang, N. Melosh, J. Feng, B. F. Chmelka, G. D. Stucky, Adv. Mater. 1998, **10**, 1380.

[8] H. Miyata, T. Noma, M. Watanabe, K. Kuroda, Chem. Mater. 2002, **14**, 766.

[9] D. Grosso, F. Babonneau, G. J. de A. A. Soler-Illia, P. A. Albouy, H. Amenitsch, Chem. Commun. 2002, 748.

[10] E. L. Crepaldi, G.J. de A.A. Soler-Illia, D. Grosso, C. Sanchez, P.-A. Albouy, Chem. Comm. 2001, **17**, 1582; D. Grosso, G.J. de A.A. Soler-Illia, F. Babonneau, C. Sanchez, P.-A. Albouy, A. Brunet-Bruneau, A.R. Balkenende, Adv. Mater. 2001, **13**, 1085; E. L. Crepaldi, D. Grosso, G.J. de A.A. Soler-Illia, P.-A. Albouy, H. Amenitsch, C. Sanchez, Chem. Mater. 2002, **14**, 3316.

[11] N. Hüsing, B. Launay, D. Doshi, G. Kickelbick, Chem. Mater. 2002, **14**, 2429; N. Hüsing, B. Launay, J. Bauer, G. Kickelbick, D. Doshi, J. Sol-Gel Sci. Technol. 2002, **26**, 615.

[12] N. Hüsing, B. Launay, G. Kickelbick, S. Gross, L. Armelao, G. Bottaro, M. Feth, H. Bertagnolli, F. Hofer, G. Kothleitner, Appl. Catal. 2003, **254**, 297.

[13] R. Supplit, N. Hüsing, H. Bertagnolli, M. Bauer, V. Kessler, G.A. Seisenbaeva, S. Bernstorff, S. Gross, J. Mater. Chem. 2006, **16**, 4443.

Organically Modified Monolithic Silica Aerogels

N. Hüsing and U. Schubert

Abstract Organically modified silica aerogels have been prepared by sol-gel processing of methanolic solutions of R'Si(OR)3 / Si(OR)4 mixtures (R = 3-methacryloxypropyl or 2-aminoethyl-3-aminopropyl groups), followed by supercritical drying with carbon dioxide. The resulting material is characterized by a very low density (0.14–0.27 g·cm^{-1}), high porosity and surface areas from 250 to 600 m^2·g^{-1} depending on the kind of organic moiety used. The organic groups are easily accessible for further reactions.

Classification

form:	porous, amorphous monoliths
function:	catalyst support, low-k dielectric, sound and heat insulation
preparation:	sol gel processing, supercritical drying
composition:	organically modified SiO_2

Introduction

Silica aerogels are highly porous and low density materials with extraordinary properties such as high specific surface areas (up to 1000 m^2·g^{-1}), good heat and sound insulation properties or transparency. Interesting technical applications as insulating materials, catalysts, sensors etc. originate from these properties.[1] The spectrum of properties and applications is expanded by incorporating functional or non-functional organic groups into the aerogels.[2,3]

The described procedure allows the synthesis of aerogels in which the organic groups are covalently bonded to the inorganic gel network. Leaching of the organic groups during synthesis, aging or drying is thus avoided. The aerogels are prepared by sol-gel processing of methanolic solutions of $R'Si(OR)_3/Si(OR)_4$ mixtures (R = Me, Et; R' = functional or non-functional organic group), followed by supercritical drying of the wet gels. During base-catalyzed sol-gel processing, the silica network forms first, and only in the second step of the reaction do the $R'SiO_{3/2}$ moieties condense onto the silicate network. The basic aerogel structure is thus retained despite the presence of the organic groups, which cover the inner surface of the SiO_2 network. They are therefore very well accessible for further reactions.

Three important variables control the composition and properties of the organically modified aerogels:

(1) *Aerogel density*: Since only little shrinkage occurs during gelation, aging and drying; the final density of the obtained material corresponds roughly to the volume of the precursor solution that is controlled by the amount of methanol (methanol is used preferentially because it is the parent alcohol of the alkoxides used in this procedure). The theoretical density of the gel body is calculated according to the formula

$$\rho_{theo} = \frac{(1-x)\, M_{SiO_2} + x\, M_{R'SiO_{3/2}}}{V_{Si(OR)_4} + V_{RSi(OR)_3} + V_{H_2O} + V_{MeOH}}$$

with ρ_{theo} = calculated density of the aerogel, x = mol % of $R'Si(OR)_3$ (setting $Si(OR)_4$ to 100%), $M_{R'SiO3/2}$ = molecular mass of the condensed form of $R'Si(OR)_3$, M_{SiO2} = molecular mass of SiO_2. Knowing the desired theoretical density, the ratios of the alkoxysilanes and the amount of water which is directly related to the portion of alkoxysilane and the total volume ($V_{(SiOR)4}$ + $V_{R'Si(OR)3}$ + V_{H2O} + V_{MeOH}), the volume of methanol which is needed to get this density can be calculated. Shrinkage during aging and supercritical drying is in the range of 5–15 % (however, higher portions of $R'Si(OR)_3$ lead to higher shrinkage >40 % due to a less connected network). Therefore, the found density of the aerogel body is slightly higher than calculated.

The density of the final aerogel can be varied from 0.070 $g \cdot cm^{-3}$ to 0.250 $g \cdot cm^{-3}$ following the given procedures. Only the volume of methanol has to be adjusted for a given $R'Si(OR)_3/Si(OR)_4$ ratio. The gel time of the system will change, but otherwise the procedures need not be modified.

(2) *Type of organic groups R'*: The given procedures are very general with regard to R'. Procedure A can be applied for all groups R' having no or only weakly basic properties, such as alkyl, aryl or $(CH_2)_nA$ (non-basic functional group A).[2] Procedure B can be applied for all groups R' having basic properties, such as $(CH_2)_nNR_2$.[3]

(3) *Portion of R' functionalized silicon atoms*: The $R'Si(OR)_3/Si(OR)_4$ ratio in the precursor mixtures can be varied without modification of the procedure. Only the amount of aqueous ammonium hydroxide needs to be adjusted. The added amount of water must correspond to one molar equivalent per alkoxy group.

$R'Si(OR)_3$ molar amounts between 0 and 40 % result in stable gel networks. Higher amounts will not give stable gels. For the synthesis of unmodified silica aerogels, procedure A is to be followed, with 100 % $Si(OMe)_4$ as the precursor. Use of more than 10 mol% of $R'Si(OR)_3$ results in incomplete incorporation of the $R'SiO_{3/2}$ groups into the aerogel network, i.e. the final aerogel then contains less R' groups than employed in the precursor mixture.

Materials

- Tetramethoxysilane (TMOS) purchased from Fluka, purity >98 %, used as received.
- 3-Methacryloxypropyltrimethoxysilane (MEMO) purchased from Fluka, purity >98%, used as received.
- 2-Aminoethyl-3-aminopropyltrimethoxysilane purchased from Wacker, purity >99 %, used as received.
- Methanol purchased from LOBA Chemie, purity 99.8 %, used as received.
- 0.01 n Ammonium hydroxide, prepared by diluting the NH_4OH volumetric standard 0.1 M (Aldrich) with deionized water.
- Pressurized carbon dioxide (60 bar) supplied by Messer Griesheim.
- Cylindrical polyethylene vials with caps on both ends (13 mm in diameter and 45 mm length. The size is determined by the dimensions of the autoclave pressure chamber used. The use of any other autoclave with any other sample dimensions is in principle possible).
- Autoclave Polaron 3000, internal dimensions of the pressure chamber are 32 mm in diameter and 75 mm in length.

Safety and Disposal

Safety and handling instructions for the chemicals are found in the corresponding materials safety data sheets (MSDS). The employed silanes may react vigorously with water. The described experiments should only be carried out in well-ventilated areas, since there is danger of severe eye damage. Standard high-pressure safety precautions are required for operating the autoclave.

Procedures

A. Preparation of 3-Methacryloxypropyl-substituted Silica Aerogel

Sol-gel processing. To an amount of 19.3 g (18.70 ml, 125 mmol) of tetramethoxysilane, 3.48 g (3.33 ml, 14 mmol) of 3-methacryloxypropyltrimethoxysilane and 14.60 g (18.48 ml, 456 mmol) of methanol are added in a 100 ml glass flask at room temperature. An amount of 10.53 g (10.53 ml, 585 mmol) of 0.01 M aqueous ammonium hydroxide is added to the solution with a syringe as quickly as possible, and the mixture is magnetically stirred for 5 min. Ammonia serves as a catalyst for sol-gel processing. The amount of water corresponds to one molar equivalent per alkoxy group (4×125 mmol for the tetraalkoxysilane + 3×14 mmol for the trialkoxysilane). The sol is transferred into the cylindrical polyethylene vials, which are then closed at both ends. The gels are kept in the solvent for seven days at 30°C. During this period it gels (for this composition after approximately 40 min). The containers have to be closed tightly to avoid drying the wet gels due to solvent evaporation.

Supercritical drying. The sample holder for the autoclave (Fig. 1) must be filled with methanol (the solvent used for the preparation of the gel) to avoid (partial) drying of the wet gel. The autoclave is cooled to 10 °C. The cylindrical vials are then opened on both sides, and the monolithic gel body is carefully pushed into the boat-shaped sample holder of the autoclave to avoid fragmenting the sample. The specimen boat is loaded into the pressure chamber, the door of the autoclave is closed, and then the *fill valve* on the pressure vessel is carefully opened. The autoclave fills with liquid CO_2, and the pressure increases to the bottle pressure (filling the autoclave with liquid carbon dioxide should be rapid). With inlet valve kept open, the *drain valve* is opened slowly to a flow of maximum 0.5 ml/min to substitute the methanol mother liquid for liquid carbon dioxide. This flushing process takes 4 d to exchange the solvents completely. After this process, all valves are closed and the actual supercritical drying process is performed.

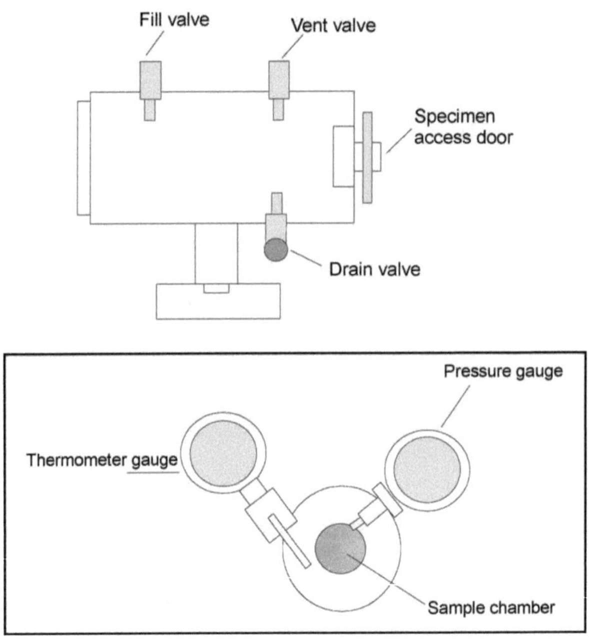

Fig. 1. Schematic diagram for autoclave with (left) side view and (right) cross

The temperature of the autoclave is raised 1° / 5 min to 40°C, while the pressure is carefully monitored and manually adjusted to a value of 100–120 bar (depending on the stability of the emergency burst disc). After reaching the supercritical state (31.5°C, 75 bar) the carbon dioxide fluid is slowly released from the autoclave. For this purpose, the *drain valve* is opened again. This process should be performed slowly, over a period of 12 h to avoid condensation effects. After reaching ambient pressure the monolithic, crack-free aerogel body is removed from the autoclave (The vent valve is only needed in case of an emergency).

Brief autoclave profile:
Filling with liquid carbon dioxide: p = 60 bar, T = 10 °C, holding time: 2–3 h
Flushing process: p = 60 bar, T = 10 °C, flow: 0.2 ml·min^{-1}, 4 d
Heating to the supercritical state: heating rate: 1°C/ 5 min, pressure at the end has to be adjusted manually to ~100–120 bar
Releasing the fluid: duration: 12 h

Characterization

The aerogel is translucent. Shrinkage during supercritical drying: 9.1±2.5 %. Final density: 0.27±0.01 g·cm^{-3}. The density is calculated from the dimensions and the weight of the monolithic gel body after supercritical drying, and the shrinkage from the difference of the dimensions before and after supercritical drying.

IR (slice of 0.5–1.0 mm thickness): 3113 (w, $\nu_{=CH}$), 1713 (m, $\nu_{C=O}$), 1639 (m, $\nu_{C=C}$).

Elemental analysis: C, 11.9; H, 1.9; Si, 36.6 (the sample is dried at 60 °C *in vacuo* prior to C,H analysis, and at 150 °C for Si analysis). C=C content: 1.02 mmol·g^{-1} (the carbon-carbon double bond can be quantified following a procedure of Byrne and Johnson).[4]

Specific surface area: 577±50 m^{2}g^{-1} (determined with a Sorptomat ASAP 2000 / Micromeritics by N_2 sorption and five-point BET analysis. The sample was dried at 110°C / 1.33·10^{-5} bar for 16 h prior to analysis).

Comments

(1) In this procedure, TMOS cannot be replaced by the less harmful tetraethoxysilane (TEOS). Different miscibilities and hydrolysis and condensation kinetics will give an aerogel with totally different properties.
(2) The surface analysis of aerogels by nitrogen sorption is not trivial. Due to the delicate network structure aerogels can be compressed during a nitrogen sorption experiment which results e.g. in an underestimation of the pore volumes.[5] Therefore, all the data from nitrogen sorption experiments have to be interpreted very carefully.

B. Preparation of 2-Aminoethyl-3-aminopropyl-substituted Silica Aerogel

Sol-gel processing. A mixture of 10.60 g (10.27 ml, 51 mmol) of tetraethoxysilane, 2.83 g (2.80 ml, 13 mmol) of 2-aminoethyl-3-aminopropyltrimethoxysilane and 24.87 g (31.48 ml, 776 mmol) of methanol is added to a 100 ml glass flask at room temperature. An amount of 4.37 g (4.37 ml, 243 mmol) of water is added to the solution with a syringe as quickly as possible, and the mixture is magnetically stirred for 5 min. The amount of water corresponds to one molar equivalent per alkoxy group (4×51 mmol for the tetraalkoxysilane + 3×13 mmol for the trialkoxysilane). The sol is transferred to the cylindrical polyethylene vial that is then closed at both ends. The gels are kept in the solvent for seven days at 30°C. During this period it gels (for this composition after approximately 30 min). The resulting wet gel is further processed as in procedure A.

Supercritical drying. Drying is performed as described in procedure A.

Characterization

The aerogel is opaque. Shrinkage during supercritical drying: 11.6±0.5 %. Final density: 0.14±0.01 $g{\cdot}cm^{-3}$. The density is calculated from the dimensions and the weight of the monolithic gel body after supercritical drying, and the shrinkage from the difference of the dimensions before and after supercritical drying.

Elemental analysis: C, 9.1; H, 4.7; N, 3.8; Si, 34.2 (for C,H,N determination the sample is dried at 60°C in vacuo prior to analysis, and for Si analysis at 150 °C).

Specific surface area: 292±100 m^2g^{-1} (determined with a Sorptomat ASAP 2000 / Micromeritics by N_2 sorption and 5-point BET analysis. The sample was dried at 110°C / $1.33{\cdot}10^{-5}$ bar for 16 h prior to analysis).

Comments

(1) In this procedure, no catalyst is needed, because the amino groups of 2-aminoethyl-3-aminopropyltrimethoxysilane act as an internal catalyst.

(2) The high concentration of amino groups promote rapid catalysis of the reaction such that gelation occurs much faster than in procedure A. $Si(OMe)_4$ (TMOS) is therefore exchanged by the slower reacting $Si(OEt)_4$ (TEOS) to get controlled gelation.

(3) The miscibility of TEOS, methanol and water is low. Therefore, a higher volume of methanol is used than in procedure A, which leads to aerogels with lower theoretical density. Thus, the upper limit of the accessible density range is determined by the miscibility of the precursors.

References

[1] N. Hüsing, U. Schubert, *Angew. Chem. Int. Ed.* 1998, **37**, 22.
[2] F. Schwertfeger, W. Glaubitt, U. Schubert, *J. Non-Cryst.Solids* 1992, **145**, 85. N. Hüsing, U. Schubert, K. Misof, P. Fratzl, *Chem. Mater.* 1998, **10**, 3024.
[3] N. Hüsing, U. Schubert, R. Mezei, P. Fratzl, B. Riegel, W. Kiefer, D. Kohler, W. Mader, *Chem. Mater.* 1999, **11**, 451.
[4] R. E. Byrne, Jr. *Anal. Chem.* 1956, **28**, 126.; J. B. Johnson *Z. Anal. Chem.* 1957, **154**, 58.
[5] G. W. Scherer, D. M. Smith, D. Stein, *J. Non-Cryst. Solids* 1995, **186**, 309.

Porous Silica Gel by Acid Leaching of Metakaolin

C. Belver and M. Á. Vicente

Abstract Solids, mainly composed of silica gel, were prepared by acid leaching of metakaolins. The metakaolins were prepared by calcination of a natural kaolin at different temperatures (600, 700, 800 and 900°C). The metakaolins thus prepared were treated with 6 *M* aqueous HCl at room temperature and at 90°C, varying the duration of the treatments. All metakaolins prepared by calcination at 600, 700 or 800°C showed a very similar reactivity, while that prepared by calcination at 900°C showed lower reactivity, due to a beginning of sintering. Treatments at room temperature did not alter either the structure or the properties of the metakaolins. Treatment under reflux conditions led to the leaching of most of the octahedral Al^{3+} cations. Silica-based solids were thus obtained reaching, under optimal activation conditions, high BET surface areas up to 219 m^2/g and total pore volumes up to 0.065 $cm^3\ g^{-1}$. Harsher treatments conditions (for longer periods of time) resulted in a drastic amorphisation of the final products. The solids obtained are promising adsorbents or catalyst supports.

Classification

form	porous, amorphous powders
function	catalyst, catalyst support, sorbent
preparation	calcination, acid leaching
composition	SiO_2

Introduction

Acid leaching is a treatment that allows one to improve the properties of natural clay materials. For this reason this method is usually called ‘acid activation’. It consists of the treatment of the clay material with solutions of inorganic acids and starts with the de-aggregation of clay particles and the removal of soluble mineral impurities. If the treatment is strong enough, it continues with the dissolution of the octahedral cations of the clay. Thus, the treated solids are composed of a mixture of non-attacked clay layers and of a hydrous, amorphous and partially protonated silica phase. The intensity of the treatments is determined by the nature of the clay material, the nature and concentration of the acid used, and the temperature and duration of the treatment. If the treatment is intense, silica gels with high surface area, high porosity and high acidicity are obtained, which are promising as sorbents or as catalyst supports and may be competitive in different applications with silica obtained by other methods.

Sepiolite and various smectites are the clay materials most often used in acid activation studies. Kaolin, a clay ore mainly composed of kaolinite, is largely used all over the world for a large variety of applications, such as ceramics, paper coating, paper filler, paint extender, rubber filler, plastic filler, cracking catalysts or cements. The chemical improvement of the properties of natural kaolins is difficult because of the high inertness of this material. That is, it is not significantly affected by acid treatment, even under severe conditions. However, metakaolinites, metastable phases obtained by calcination of kaolinite at temperatures between 500 and 900°C, are significantly more reactive than kaolin, particularly in acid media. The transformation kaolinite–metakaolinite involves the loss of constitutional water with a reorganization of the structure. Only a small part of the Al cations remains octahedrally coordinated, while the rest is transformed into much more reactive tetra- and penta-coordinated units. The calcination temperature is an important factor for the properties of the obtained metakaolin; if a certain temperature is exceeded, the solid may sinter, resulting in a mixture of mullite and cristobalite. The optimal calcination temperature depends on the characteristics of each natural kaolin.

Differing results may be obtained when repeating this procedure, because of differences in various kaolin samples, even if obtained from the same clay deposit. This is due to small variations in the composition of the kaolin (amount and nature of impurities), to the crystallinity of kaolinite, etc.

Because of the large natural occurrence of kaolin, we describe here a systematic study of obtaining porous silica from this mineral. Natural kaolin is calcined at four different temperatures (600, 700, 800 and 900°C) to obtain metakaolins, and then the metakaolins are treated with concentrated HCl solutions to obtain porous silica.

Materials

- Kaolin from Navalacruz (Spain), purchased from Arcillas Blancas, purified by dispersion in water (no additives were needed) and decantation, extracting the column that, following the Stokes equation, corresponds to clay particle size ≤ 2μm fraction.
- Hydrochloric acid, 35%, purchased from Panreac, used as received.

Safety

Safety and handling instructions for the chemicals and for kaolin are found in the corresponding materials safety data sheets (MSDS). Standard safety precautions are required for operating with HCl solutions.

Procedures

A. Preparation of Metakaolins

The ≤ 2μm fraction of kaolin was calcined in air at 600, 700, 800 and 900°C, respectively, in a programmable furnace, with a heating rate from room temperature to the calcination temperature of 10°C min^{-1}. Once the calcination temperature had been reached, the solids were maintained isothermally for 10 h, and then cooled at 50°C min^{-1}.

B. Preparation of the Acid-activated Solids

Acid treatment was carried out with 6 *M* aqueous HCl solution at room temperature or 90°C, in the latter case under reflux conditions. For each temperature, times of treatment were 6 and 24 h. 6.0 g of the metakaolin were mixed with 180 mL of the acid solutions, and stirred for the time indicated above. After this, the suspensions were centrifuged and the solids were washed with distilled water until no chloride anions could be detected (Ag^+ test), and dried at 50°C.

Characterization

Powder X-ray diffraction (Siemens D-500 diffractometer) shows that all the metakaolins have amorphous structures, contrasting with the crystalline structure of the natural kaolin. The amorphous pattern is maintained in the acid-treated solids, although the wide band between 2θ = 20 and 40° appears, also named the halo, characteristic of amorphous silica. Chemical analysis shows a progressive leaching of Al^{3+} when the treatment is intensified, and the solids reach SiO_2 contents of 97% (referred to water-free solids), while chemical and thermal analyses show up to 25% of water content in these solids. The metakaolins have specific BET surface areas (Micromeritics ASAP 2010 adsorption analyzer) of 7–13 m^2g^{-1} and total pore volume of 0.0015 cm^3g^{-1} (18 m^2g^{-1} and 0.0007 cm^3g^{-1} in the natural kaolin). After the acid treatment these values increase to 219 m^2g^{-1} and 0.065 cm^3g^{-1}, respectively. FT-IR spectroscopy confirms the transformation of the clay structure to amorphous silica by the shift of the Si– O characteristic band from 1000 to 1100 cm^{-1}. The concentration of Brönsted acid sites in the activated solids (retention of cyclohexylamine, determined by thermogravimetry) reaches 0.19 mmol g^{-1}. VT-DRIFTS spectra (Mattson Polaris FTIR spectrometer with a Graseby Selector DRIFTS accessory combined with an environmental chamber, purged continuously with nitrogen gas) collected from 200 to 500°C at 50°C increments, showed that the area under the broad OH stretching band decreased as the sample temperature was increased.[1-6]

Comments

(1) Thermal treatment transforms natural kaolin into metakaolins in an efficient way. The transformation is easily followed by the changes in the XRD patterns, from the very intense kaolin peaks to the amorphous structure of the metakaolins. The surface area decreases during the calcination. The metakaolins prepared by calcination at 600, 700 and 800°C have very similar properties and are much more reactive than the parent kaolin, being modified in an effective way by the acid treatments. However, the 900 °C-synthesized metakaolin is much less reactive, probably due to the beginning of a sintering process at this temperature. At high temperature, metakaolins are transformed into a mixture of mullite and silica, the temperature of such transformations is dependant on the nature, mineralogical impurities, etc. of each natural kaolin.

(2) The attack to the structure of the metakaolins can be easily followed by PXRD by the increase in the intensity of the wide band between $2\theta = 20$ and 40° typical of silica. It can also be followed by the progressive shift of the SiO band in the FTIR spectra, from 1000 cm^{-1} (typical of silicates and due to Si–O–Al–O–Si bonds) to 1090 cm^{-1} (typical of silica, and due to Si–O–Si–O–Si bonds).

(3) Acid activation with 6 *M* HCl clearly modifies the structure and properties of the metakaolins. Acid treatments with 1 *M* HCl did not modify the metakaolins, independent of the time or temperature of the treatments.

(4) Al^{3+} octahedral cations are progressively removed by the acid solutions. The degree of dissolution depends on the intensity of the treatment and is higher when reflux conditions are used and when the time of the treatment was 24 h instead of 6 h. Up to 95% of the octahedral cations are removed, thus producing solids composed almost exclusively by silica, with a high degree of hydration (up to 25%).

(5) The BET surface area of the solids develops during activation, reaching relatively high values under optimal conditions, *viz.* 171–219 $m^2\ g^{-1}$ for the metakaolins synthesized by calcination at 600, 700 and 800°C and treated with 6 *M* HCl for 6 h under reflux (total pore volume up to 0.065 $cm^3\ g^{-1}$). However, more intense treatment, especially when the duration of the treatment is prolonged to 24 h, has a deleterious effect on the surface area, which then decreases to ca. 22 $m^2\ g^{-1}$ (total pore volume $\approx$ 0.001 $cm^3\ g^{-1}$). The decrease in

the surface area is due to an 'amorphization' of the silica generated, provoked by the very fast alteration occurring under these conditions. It was observed that the amorphization is a rather fast process when critical conditions are exceeded. The acidity of the treated solids also depends on the intensity of the treatment, reaching values up to 0.19 mmol of Brönsted sites per gram.

(6) The VT-DRIFTS study of the acid-activated metakaolin during 6 h suggests that water molecules experience a wide range of hydrogen bond strengths when bound to the activated metakaolin surface. As the temperature was increased, the range of hydrogen bond strengths and the number of water molecules bound decreased, resulting in a weaker and narrower OH stretching band. Above 300ºC it was possible to fit this broad OH band to seven contributing components at 3731, 3699, 3655, 3615, 3583, 3424 and 3325 cm^{-1}. As the temperature was increased, the 3731 cm^{-1} band (silanol groups) increased in relative intensity due to the removal of water molecules. All the other bands decreased in intensity as the water molecules and OH groups were thermally desorbed.

(7) In summary, acid activation is an effective method for the improvement of the properties of kaolins. An intermediate step (calcination to form metakaolins) is needed, and concentrated acid solutions must be used, and the development of textural properties is not as high as in smectites. Despite these disadvantages, the great natural occurrence of kaolin may justify the preparation of porous silica from this mineral. The strongly acidic solids, with BET surface area of *ca.* 220 m^2g^{-1}, are promising materials for catalytic supports and sorbents.

References

[1] M. A. Vicente. C. Belver, R. Trujillano, M. Suárez, M.A. Bañares, V. Rives. In: Applied Study of Cultural Heritage and Clays, Eds. J.L. Pérez Rodríguez. Consejo Superior de Investigaciones Científicas, Madrid, 2003, pp. 519–534.

[2] C. Belver, M.A. Bañares, M.A. Vicente. Chem. Mater. 2002, **14**, 2033.

[3] C. Belver, M.A. Bañares, M.A. Vicente. Stud. Surf. Sci. Catal. 2002, **144**, 307.

[4] C. Belver, M.A. Bañares, M.A. Vicente. Bol. Soc. Esp. Ceram. V. 2003, **43**; 148.

[5] C. Belver, Ph.D. Thesis, University of Salamanca, 2004.

[6] C. Belver, C. Breen, F. Clegg, C. E. Fernandes, M. A. Vicente, Langmuir, 2005, **21**, 2129.

the surface area is due to an 'amorphization' of the silica generator, provoked by the very fast alteration occurring under these conditions. It was observed that the amorphization is a rather fast process when critical conditions are exceeded. The acidity of the treated solids also depends on the intensity of the treatment, reaching values up to 0.19 mmol/g Brønsted sites per gram.

The FTIR study of the acid-activated metakaolin [illegible] shows that [illegible] experienced a wide [illegible] of hydrogen [illegible] [illegible]

Zirconia-Pillared Clays

M. A. Vicente and A. Gil

Abstract Zirconia pillared clays were prepared by intercalation of natural montmorillonite and natural saponites with zirconium polyoxo cations (prepared either by partial hydrolysis of zirconyl chloride or from zirconium acetate solution), followed by calcination. The use of the Zr acetate solution is a much less aggressive method. Well-ordered layered solids were mostly obtained, while delamination and acid leaching were observed to happen, together with intercalation, in other cases. The resulting materials contain high amounts of ZrO_2, and have high surface areas (~300 m^2g^{-1}) and pore volumes (~0.200 cm^3g^{-1}).

Classification

form:	porous powders
function:	molecular sieve, adsorbent, catalyst, catalyst support
preparation:	intercalation, ion exchange, pillaring
composition:	ZrO_2-modified clay

Introduction

Pillared interlayered clays (in short PILCs) form one of the most interesting families among the solids prepared by molecular engineering. These solids are prepared by exchange of the charge-compensating cations of smectitic clays by inorganic polyoxo cations, followed by calcination. The first process is usually denoted to as "intercalation" and the second as "pillaring"; sometimes the total process is named "pillaring". During intercalation, the clay layers are separated to a distance larger than that in natural clays, because of the much larger height of the pillaring polyoxo metal cations compared with the naturally occurring charge-compensating cations. During the preparative procedure, the intercalated polyoxo metal cations are transformed to oxo-hydroxo phases, the "pillars", by dehydration and dehydroxylation reactions that keep the clay layers separated to a distance much larger than in natural samples. In general, all smectitic clays can be pillared, although parent clays with high crystallinity and layer charge close to 1.0 meq/g usually give rise to pillared solids with better properties, in terms of basal spacing, crystalline ordering and surface area and porosity.

Pillared clays are stable because of the high stability of the inorganic pillars. A molecular-size two-dimensional porous network is thus created between the horizontal clay layers and the vertical pillars that avoid their collapse. This network is characterized by the density of pillars, their size and height, etc., factors conditioned by the nature of the original clay and by the nature of the intercalating polyoxo metal cations. The porous structure and the physicochemical properties of the pillared clays can be tailored, to a certain point, by controlling these parameters. An alternative view on these solids is that the clay layers avoid the aggregation of the polyoxo metal cations during calcination, thus giving rise to a highly dispersed phase of the oxo-hydroxo compound used in the intercalation reaction.[1]

To be useful for intercalation experiments, an element should form stable and structurally well defined polyoxo cations upon partial hydrolysis. Al(III) easily forms $[Al_{13}O_4(OH)_{24}(H_2O)_{12}]^{7+}$ under well-established conditions. Thus, the intercalation of this polycation into clays is also established, and Al-pillared clays are well documented in the literature.[1] Scheme 1 shows the general pillaring process, which is similar for pillaring with other elements.

The interest in pillaring with zirconia species is evident from the excellent properties of zirconia. The formation of the $[Zr_4(OH)_8(H_2O)_{16}]^{8+}$ polycation upon hydrolysis of Zr(IV) salts, mainly zirconyl chloride, has been reported. However, because of the strongly acidic character of this cation, the polymerization must to be carried out under very acidic conditions, which may strongly affect the clay structure during intercalation. In this report, the synthesis of zirconia-pillared clays under soft conditions is described, starting from zirconium acetate solutions. This method is compared with the usual zirconyl chloride route.

Materials

- Saponite from Ballarat (USA), purchased from The Clay Minerals Repository, purified by dispersion-decantation (dispersion of the clay in water and further decantation of the suspension, extracting the upper layer of the suspension that, by application of Stokes law, contains the clay particles, i.e. the particles ≤ 2 μm).
- Saponite from Yunclillos (Spain), purchased from TOLSA, purified by dispersion-decantation.
- Montmorillonite from Minas de Gador (Spain), purified by dispersion-decantation.
- Zirconyl chloride, $ZrOCl_2.8H_2O$, 99.9+%, purchased from Sigma-Aldrich, used as received.
- Zirconium acetate solution in dilute acetic acid, containing 15-16 wt% of Zr (reagent 7585-20-8 from Aldrich) , used as received.

- Hydrochloric acid, 35%, purchased from Panreac, used as received.

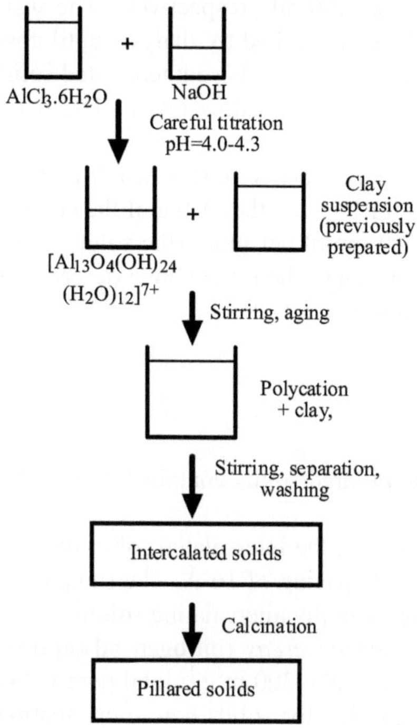

Scheme 1. General pillaring process

Safety and Disposal

Safety and handling instructions for the chemicals and for the clays are found in the corresponding materials safety data sheets (MSDS). Zirconyl chloride is corrosive and hygroscopic. Zirconium acetate solutions are harmful and irritant. Standard safety precautions are required for operating with HCl solutions.

Procedures

A. Preparation of Zr-pillared Clays Using Zirconyl Chloride as Precursor

The same procedure was used for pillaring the three clays. 8.0 g of a clay were added to freshly prepared 0.1 M solutions of $ZrOCl_2{\cdot}8\ H_2O$ in water, and stirred for 2 h. Two Zr/clay ratios in the intercalating solutions were considered, *viz*. 3.0 and 20.0 mmol Zr per gram. This corresponds to 7.7 g of zirconyl chloride in 240 mL of water, or 51.6 g in 1600 mL, respectively. The suspensions were then centrifuged, and the solids were washed by dialysis until absence of chloride anions (4 d), and dried at 50°C overnight. These intercalated solids (Series A1) were calcined at 200 and 400°C for 4 h (heating rate of 1°C per minute), giving the pillared solids.

In a second series of experiments (Series A2), 0.1 M solutions of $ZrOCl_2.8H_2O$ in water were refluxed for 2 h, and then 8.0 g of the clays were added. In this case, the Zr/clay ratio was 3.0 mmol per gram (the solutions contain 7.7 g of zirconyl chloride in 240 mL of water). Then, the above described procedure was used for the preparation of the solids.

Characterization

Chemical analysis: the pillared solids contain 10–16% (Series A1) or 25–28% of ZrO_2 (Series A2).

Powder X-ray diffraction: the intercalated solids of Series A1 maintain the layered structure, with basal spacing of 16 Å. The samples of Series A2 completely delaminate after contact with the intercalating solution.

Specific Langmuir surface areas (nitrogen adsorption; Micromeritics ASAP 2010 analyzer): Series A1: 203–300 m^2g^{-1}; total pore volume 0.143–0.220 cm^3g^{-1}, depending on the parent clay for solids before calcination; decrease by 20–30% upon calcination. Series A2: surface area of 263–348 m^2g^{-1}, and pore volume of 0.105–0.158 cm^3g^{-1}.

Concentration of Brönsted acid centers in the pillared solids of the A1 series (retention of cyclohexylamine, determined by thermogravimetry) is of 0.46–0.87 mmol g^{-1}.[2-7]

Comments

This procedure allows to an efficient incorporation of Zr-species into the clay, but it is highly aggressive for the clay structure. The pH of the intercalating solutions was 0.7–1.1, and this caused acid leaching of octahedrally coordinated cations of the clays, especially Mg^{2+}. Magnesium cations occupy the majority of octahedral sites in the saponites, and are present in an important amount in the montmorillonite. The non-reflux intercalated solids have well-ordered layered structures, which, however, collapse very easily. Complete collapse is observed upon calcination at 200ºC. The solids treated with the refluxed intercalated solutions are completely delaminated, i.e., the Zr-species generated under reflux are able to separate the clay layers completely from each other. The delaminated solids have, nevertheless, interesting properties.

B. Preparation of Zr-pillared Clays Using Zirconium Acetate as Precursor

The intercalating solutions were prepared by diluting the needed amounts of the commercial solution, corresponding to an average Zr content of 15.5%, in 175 mL of water. Two Zr/clay ratios were considered, 3.0 and 20.0 mmol per gram, for which 11.4 and 76.1 mL of the original solution, respectively, were required. Then, 8.0 g of the clays were added to the solutions, which were then stirred vigorously for 2 h. After this period, the suspensions were centrifuged, and the solids washed by dialysis until absence of chloride and acetate (4 d), and dried at 50°C overnight. These solids were calcined at 200, 400 and 500 °C for 4 h (heating rate of 1 °C per minute), giving the pillared solids.

Characterization

Chemical analysis: the pillared solids contain 20–36% of ZrO_2.

Powder X-ray diffraction: the intercalated solids maintain the layered structure, with basal spacing of 16–21 Å. The layered structure is retained after calcination at 200 °C, but collapses when the solids are calcined at 400°C.

FT-IR spectroscopy: the acetate groups remain bonded to the Zr atoms as bidentate ligands in the intercalated solids.

Thermal analyses indicate that the removal of these organic moieties happens in two steps between 300-400°C, and is always completed at 400°C.

Specific Langmuir surface areas: 131–282 m^2g^{-1}, depending on the parent clay used; increase to 299–324 m^2g^{-1} in the pillared solids. Total pore volume: 0.108–0.178 cm^3g^{-1} in the intercalated solids; increase to 0.191–0.213 cm^3g^{-1} in the pillared solids.

Concentration of Brönsted acid centers in the pillared solids: 0.39–0.66 mmol g^{-1}.[2-7]

Comments

This procedure allows preparing Zr-pillared clays by a method less aggressive to the clays. The pH of the intercalating solutions is 3.3. The polymeric Zr-species, containing organic ligands, are intercalated into the clay layers in an efficient way, substituting the exchangeable cations, but without affecting the octahedral or tetrahedral clay layers. The intercalated solids show well-ordered layer structures, which depend on the nature, especially in the crystallinity, of the parent clay. The highly crystalline Ballarat saponite gives rise to very well ordered intercalated solids, while Gador montmorillonite, poorly ordered itself in the natural form, yields less ordered solids. The organic moieties of the intercalating species blocks access to the pore system of the intercalated solids, but this access is opened upon calcination. The acetate groups are removed in two steps, separated by about 10º. The removal of these ligands also depends on the nature of the clays, their acidity probably catalyzing the thermal degradation of the ligands, which is complete at 330-390ºC. The structures are stable up to 400ºC, and solids with high surface areas and porosities, and a large proportion of acidic centers are obtained. The latter renders them promising as molecular sieves, catalysts and catalytic metal supports.

References

[1] A. Gil, L.M. Gandía, M. A. Vicente, Catal. Rev.-Sci. Eng. 2000, **42**, 145.
[2] R. Toranzo, M. A. Vicente, M. A. Bañares-Muñoz, L. M. Gandía, A. Gil, Micropor. Mesopor. Mater. 1998, **24**, 173.
[3] L. M. Gandía, R. Toranzo, M. A. Vicente, A. Gil, Appl. Catal. A 1999, **183**, 23.
[4] A. Gil, M. A. Vicente, L. M. Gandía, Micropor. Mesopor. Mater. 2000, **34**, 115.
[5] L. M. Gandía, A. Gil, M. A. Vicente, Appl. Catal. A 2000, **196**, 281.
[6] A. Gil, M. A. Vicente, L. M. Gandía, Bol. Soc. Esp. Cer. Vid. 2000, **39**, 530.
[7] M. A. Vicente, M. A. Bañares-Muñoz, L. M. Gandía, A. Gil, Appl. Catal. A 2001, **217**, 191.

Montmorillonites with Mixed Aluminum-Lanthanide Oxide Pillars

C. Pesquera, C. Blanco and F. González

Abstract Pillared montmorillonites with mixed aluminum/cerium or aluminum/lanthanum pillars were prepared. Nuclear magnetic resonance studies indicated absence of tetrahedral aluminum in the pillars. The materials have a high thermal stability, and high specific surface area and porosity, with pores at the limit between micropore and mesopore sizes. The number and strength of the acid sites in these materials is also high.

Classification

form:	porous powders
function:	catalyst support, acid catalyst.
preparation:	intercalation, ion exchange, pillaring
composition:	montmorillonite modified with polyoxocations

Introduction

Pillared clays in general and pillared montmorillonites in particular are clay minerals that have been modified by introducing large polyoxycations into their interlayer regions. The separation between layers depends on the polyoxycation used and can be kept stable. Heating these materials results in the formation of inorganic oxide clusters that prop the clay layers permanently, thus generating a microporous structure with a high specific surface area. The inorganic polyoxycations most frequently used as pillaring agents are aluminum, zirconium, titanium, chromium and iron-containing species.[1-5]

Since the introduction of pillared clays in the late seventies,[6] much research has been done to develop materials suitable as active components in catalysis and for other applications.[7] Pillared clays are versatile materials, since the size and shape of their cavities can be varied over wide ranges, and the constitution and chemical properties of the pillars can be altered. Pillared montmorillonites have been proposed as potential materials for cracking catalysts of heavy oil fractions,[7] as they can be prepared with pore sizes larger than those of zeolites. The presence of acid centers on the surface of their layers, as well as on their pillars, makes these materials suitable also for use in other reactions of acid catalysis.[8]

In order to prevent the clay layers from sintering at the high temperature of the catalytic reactions, the stability of the pillars must be increased. One way to achieve this is to introduce mixed-metal pillars into the materials.[9-12] It was found that the incorporation of lanthanide elements resulted in materials whose basal spacing was larger than in conventional pillared materials.[13-16]

Materials

- Montmorillonite: natural clays from different mineral deposits. Specific surface area (BET) = 33—80 m^2/g; cation exchange capacity between 83 and 110 mequiv/100 g.
- Basic aluminum chloride [$Al_2(OH)_5Cl_2 \cdot 3H_2O$], purchased from Hoechst, purity 99,5%.
- $LaCl_3 \cdot 7H_2O$, purchased from Merck, purity 99%.
- $CeCl_3 \cdot 7H_2O$, purchased from Merck, purity 98.5%
- Cylindrical Teflon vial with caps (25 mm in diameter and 250 mm in length). This size is determined by the dimensions of the autoclave pressure chamber used.

Safety and Disposal

Safety and handling instructions for the chemicals are found in the corresponding materials safety data sheets (MSDS). Standard high pressure safety precautions are required for operating the autoclave.

Procedures

Preparation of the Pillaring Agent

The solutions of the pillaring agent with molar ratio Al/Ce = 25 were prepared by adding 0.49g of $CeCl_3 \cdot 7H_2O$ to a solution of 7.2 g of [$Al_2(OH)_5Cl_2 \cdot 3H_2O$] in 21.6 mL of water. The solutions of the pillaring agent with ratio Al/La = 25 were prepared by adding 0.48g of $LaCl_3 \cdot 7H_2O$ to a solution of 7.2 g of [$Al_2(OH)_5Cl_2 \cdot 3H_2O$] with 21.6 mL of water .The mixture was magnetically stirred for 15 min. The best properties are obtained for a ratio Al/lanthanide of 25.

The final solutions, 2.5 M in Al, were transferred to a cylindrical Teflon vial that was then closed, placed in an autoclave and heated to 130°C at 10°C/min and maintained at this temperature for 72 h. After returning to room temperature and atmospheric pressure, the reaction mixture was diluted with 640 mL of water necessary to yield an Al concentration of 0.1 M.

Pillaring Process

The solutions of pillaring agent were added slowly from a glass beaker, with vigorous stirring, to a clay slurry obtained by mixing 10 g of clay in 400 mL of water which had been left to stand for 72 h. The final proportion in all cases was 20 mequiv of Al/g of clay, with a solid/liquid ratio of 0.5%. The reaction mixture was stirred continuously for 24 h at room temperature. It was then dialyzed with distilled water, using 1 L of water/g of clay. Dialysis was continued, with water being renewed every 24 h until the Cl^- concentrations decreased to the point where the conductivity of the aqueous phase was <30 μS. Finally, the samples were freeze-dried. This process consists in the sublimation of water at low pressure after previous solidification at liquid nitrogen temperature. Samples were then calcined in air atmosphere for 2 h at 400°C. The obtained samples are denoted as AlCe-PILC and AlLa-PILC.

Characterization

The amount of aluminum incorporated (mequiv/g clay) in the pillared materials is 15.6 mequiv Al/g clay for AlCe-PILC, and 19.5 mequiv Al/g clay for AlLa-PILC, which represents 78–97.5% of the aluminum essayed, whereas the incorporated cerium or lanthanum is 0.3 mequiv/g clay (37.5%). The cation proportions were determined by energy dispersive X-ray spectrometry (EDS).

X-ray diffraction (Cu K_α radiation): diagrams in the range 3–12° (2θ) display a d(001) peak corresponding to spacings between the clay layers of 27.2 Å for AlCe-PILC and at 27.0 Å for AlLa-PILC before the calcination process. In the samples treated at 400°C, the spacings were reduced to 24.7 Å and 26.1 Å respectively.

^{27}Al MAS NMR spectroscopy (78.23 MHz, pulse width of 4 μS, spinning rate 3.0 kHz, spectral width 50 kHz): Chemical shifts are relative to 0.1 M $[Al(H_2O)_6]^{3+}$: Spectra of the aluminum/lanthanide-pillared samples display a central line close to 0 ppm, assigned to the hexa-coordinate Al of the octahedral layer of montmorillonite. The spectrum for montmorillonite shows a very weak signal around 60 ppm, which is attributed to tetrahedral Al replacing Si in the tetrahedral layer. A sample pillared only with aluminum (Al-PILC) shows a signal of higher intensity at 60 ppm due to the presence of tetrahedral aluminum in the pillars. In contrast, this signal is not altered with respect to the initial montmorillonite in the

aluminum/lanthanide-pillared samples. This indicates the absence of tetrahedral aluminum in the intercalated inorganic polyoxycations between the clay sheets. The ^{27}Al MAS NMR spectra of Al-PILC and AlLa-PILC are shown in Fig. 1.

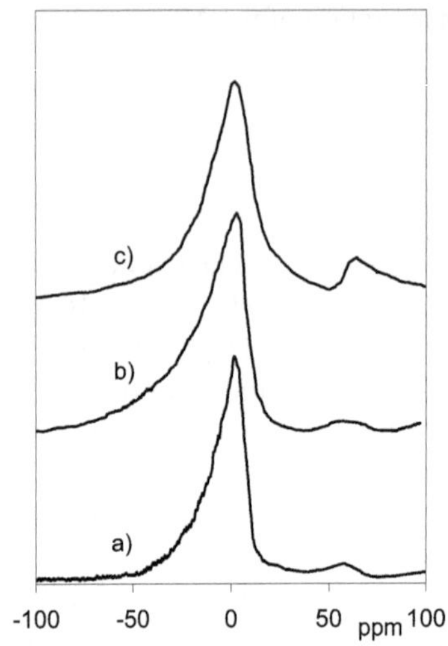

Fig. 1. ^{27}Al MAS NMR spectra: a) montmorillonite; b) AlLa-PILC; c) Al-PILC.

Nitrogen adsorption isotherms of the pillared materials show a great increase in the volume of adsorbed nitrogen around $p/p_o = 0.1$, which is at the border between micropore and mesopore sizes. Fig. 2 shows the pore size distribution obtained for pillared samples. For Al-PILC, the micropore volume accumulates around a single pore type. In contrast, the pore volume distribution for the Al/lanthanide-pillared samples shows two maxima.

The specific surface area, determined by nitrogen adsorption and BET analysis, is 350 m^2g^{-1} for AlLa-PILC and 375 m^2g^{-1} for AlCe-PILC after drying the samples at 140°C and 10^{-5} bar for 20 h. 80-85% of the specific surface area is retained at 700°C.

Chemisorption of ammonia was analyzed from the adsorption-desorption isotherm in the pressure range 50–400 mmHg (6.7–53 kPa). The proportion of acidic centers obtained from ammonia chemisorbed at 200°C is 300 μmol per gram of AlLa-PILC and 350 μmol per gram of AlCe-PILC.

Comments

The aluminum/lanthanide-pillared montmorillonites have characteristics very different to that of montmorillonite pillared only with aluminum. They show a bimodal micropore structure between microporosity and mesoporosity. The textural parameters, i.e. specific surface area and micropore volume, show higher values and the materials are thermally more stable, maintaining high values of specific surface area and micropore volume up to 700ºC. These materials show more acidic centers than the sample pillared with only aluminum.

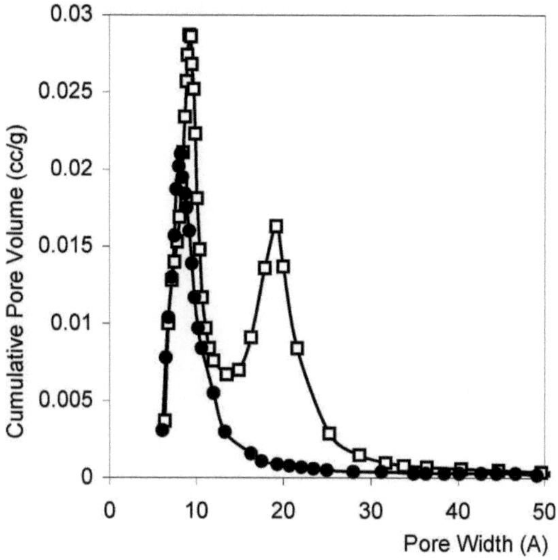

Fig. 2. Pore volume distribution of the Al-PILC (●) and AlLa-PILC (□).

References

[1] A. Gil, L. M. Gandia, M. A. Vicente, Catal. Rev. Sci. Eng. 2000, **42**,145.

[2] A. Riego, I. Herrero, C. Pesquera, C. Blanco, I. Benito, F. González, Appl. Clay Sci. 1994, **9**,189.

[3] F. Figueras, Catal. Rev. Sci. Eng. 1988, **30**, 457.

[4] M. A. Martín-Luengo, H. Martins-Carvalho, J. Ladriere, P. Grange, Clay Miner. 1989, **24**, 495.

[5] B. M. Choudary, V. I. K. Valli, J. Chem. Soc. Chem. Commun. 1990**,** 1115.

[6] D. E. W. Vaughan, R. Lussie, J. Magee, US Patent. 4,176,090 (1979).

[7] I. V. Mitchell, Pillared layered structures, current trends and applications, Elsevier Applied Science, London, 1990.

[8] E. Kikuchi, T. Matsuda, Catal. Today 1988, **2**, 297.

[9] M. L. Ocelli , J. Mol. Catal. 1986, **35**, 377.

[10] S. M. Bradley, R. A. Kydd,. Catal. Lett. 1991, **8**, 185.

[11] X. Tang, W. Q. Shu, Y. F. Shen, S. L. Suib, Chem. Mater. 1995, **7**, 102.

[12] M. J. Hernando, C. Pesquera, C. Blanco, I. Benito, F. González, Chem. Mater. 1995, **8**, 76.

[13] J. Sterte, Clays Clay Miner. 1991, **39**, 167.

[14] J. R. Mc Cauley, US Patent 4,818,737 (1988).

[15] M. J. Hernando, C. Pesquera, C. Blanco, F. González, Chem. Mater. 2001, **13**, 2154.

[16] M. J. Hernando, C. Pesquera, C. Blanco, F. González, Stud. Surf. Sci. Catal. 2002, **144**, 617.

Birnessite-type Manganese Oxide by Redox Precipitation

Q. Feng, Z.-H. Liu and K. Ooi

Abstract Preparation of a birnessite-type manganese oxide, with a layered structure, by redox and hydrothermal reactions is described. A poorly crystalline birnessite is prepared by reacting an aqueous $Mn(NO_3)_2$ solution with a NaOH / H_2O_2 solution at room temperature. A highly crystalline birnessite is obtained by hydrothermal treatment of the poorly crystalline birnessite in a NaOH solution.

Classification

form:	sub-micron crystalline powder
function:	selective adsorbent, catalyst, cathode material for secondary batteries, precursor for nanomaterials
preparation:	redox precipitation, hydrothermal treatment
composition:	$Na_{0.31}MnO_{1.91}\cdot 0.7H_2O$, $Na_{0.38}MnO_{1.95}\cdot 0.7H_2O$

Introduction

Birnessite-type manganese oxide is a layered compound with cations and water molecules in the interlayer space of negatively charged layers of edge-sharing MnO_6 octahedral (Fig. 1).[1-2] The basal spacing of birnessite is typically 0.7 nm, but further hydration can increase the spacing to produce a 1 nm birnessite known as buserite.[3-4] Birnessite shows excellent ion-exchange, adsorption, and intercalation properties for cations and molecules. These properties make birnessite a useful material in many applications, for example as a cation sieve material for separation of metal ions [5-6] and nuclear waste adsorption,[7] a molecular sieve for heterogeneous catalysis in the reduction of NO in the presence of ammonia,[8] oxidation of CO,[9] and hydrogenation of alkenes,[10] and as a cathode material for rechargeable lithium ionic batteries.[4,11-13] Birnessite is also a useful intermediate in soft chemical syntheses. Manganese oxide ionic sieves and molecular sieves with tunnel structures can be prepared from birnessite.[14-17] The layered structure of birnessite can be exfoliated into nano-sheets in an organic amine solution.[18] The nano-sheets can also be converted into other manganese oxide nanomaterials, such as nanotubes [19] and nanofibers.[20]

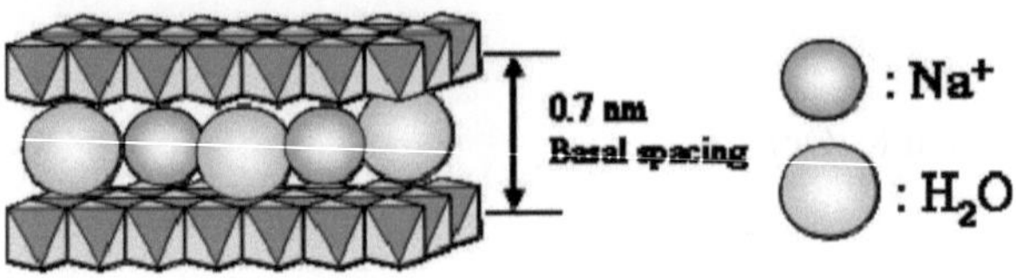

Fig. 1. Layered structure of birnessite-type manganese oxide.

Birnessite has been prepared by several synthesis processes, including passing O_2 through suspensions of $Mn(OH)_2$ obtained by reacting NaOH and Mn^{2+},[21-22] redox reactions between permanganate and Mn^{2+} in basic solution,[21,23] hydrothermal treatment of MnO_2 or Mn_2O_3 in NaOH solution,[5,24] hydrothermal treatment of $NaMnO_4$ in a slightly acidic solution (pH 3.5),[4] sol-gel processes combined with calcination,[11,25] and melting salt flux process.[26]

In this contribution a simple and convenient redox precipitation process for the synthesis of birnessite is described, which includes preparations of a poorly crystalline birnessite by redox reaction at room temperature and a highly crystalline birnessite by hydrothermal treatment of the low crystalline birnessite.[27-28]

Materials

- $Mn(NO_3)_2{\cdot}6H_2O$ (99.9%) purchased from Wako Pure Chemical Industries, Ltd, used as received.
- NaOH (96%) purchased from Wako Pure Chemical Industries, Ltd, used as received.
- H_2O_2 solution (30%) purchased from Wako Pure Chemical Industries, Ltd, used as received.
- distilled water.

Safety and Disposal

Safety and handling instruction for the chemicals are found in the corresponding materials safety data sheets (MSDS). $Mn(NO_3)_2$ solutions react with mixed solution of NaOH and H_2O_2 vigorously, with concomitant release of O_2 gas, the reaction should be carried out in a beaker, but not in a flask. Standard high-pressure safety precautions are required for operating the autoclave.

Procedures

A. Redox Precipitation

A poorly crystalline Na-birnessite with Na^+ ions in the interlayer is prepared by reacting an aqueous $Mn(NO_3)_2$ solution with a mixed solution of NaOH and H_2O_2 at room temperature. A 0.3 M $Mn(NO_3)_2$ solution is prepared by dissolving 4.3 g (15 mmol) of $Mn(NO_3)_2 \cdot 6H_2O$ in 50 mL of distilled water in a 1 L beaker. A mixed solution (100 mL) of 3% H_2O_2 and 0.6 M NaOH is prepared by dissolving 2.4 g (60 mmol) of NaOH in 90 mL of distilled water in a 200 mL beaker, and then adding 10 mL of 30% H_2O_2 solution. The $Mn(NO_3)_2$ solution is stirred as quickly as possible by hand with a glass bar, and the H_2O_2 / NaOH solution is poured into the $Mn(NO_3)_2$ solution as quickly as possible under vigorous stirring. Stirring is continued for 10 min. When the H_2O_2 / NaOH solution is poured into the $Mn(NO_3)_2$ solution, a black- brown precipitate is formed immediately, and O_2 gas is released. The product is kept in the reaction solution and aged at room temperature for 12 h. The precipitate is then filtered off, and three times washed with distilled water (100 mL). The wet sample is put in a beaker (250 mL) for freeze-drying, set on a freeze dryer, and then freeze-dried for 24 h.

Comments

The concentration of NaOH is an important parameter in this process. If the concentration is too low, β-MnOOH is formed as by-product. The amount of by-product decreases, but the crystallinity of birnessite also decreases when the NaOH concentration is increased. Birnessite is formed in a two-step reaction. In the first step $Mn(OH)_2$ is formed by reaction of Mn^{2+} with OH^-, and in the second step $Mn(OH)_2$ is oxidized to birnessite. Insertion of Na^+ and H_2O into the interlayer of the birnessite structure accompanies oxidation. When the NaOH concentration is too low, $Mn(OH)_2$ cannot be completely oxidized to birnessite, but instead to β-MnOOH, and β-MnOOH is difficult to be oxidized to birnessite under the room temperature conditions. Since manganese species are catalysts for the decomposition of H_2O_2, it is important to avoid contamination of the H_2O_2 / NaOH solution by manganese species before the reaction.

A-birnessites with A^+ cations in the interlayer (A = Li, K, Rb, or Cs) can be prepared by using the process described above with AOH solutions instead of NaOH solution. Since Li-birnessite is unstable in LiOH solution and transforms easily to a spinel-type lithium manganese oxide, aging is carried out at room temperature for 1 h for the preparation of Li-birnessite.[29]

B. Hydrothermal Treatment

A highly crystalline Na-birnessite is prepared by hydrothermal treatment of the poorly crystalline Na-birnessite prepared above using a Teflon-lined, sealed, stainless steel autoclave (Fig. 2a). The internal volume of the pressure chamber is 30 mL. The autoclave can be fixed on a rotation shaft in an electric oven (Fig. 2b). The sample in the autoclave is stirred by rotating the shaft during the hydrothermal treatment.

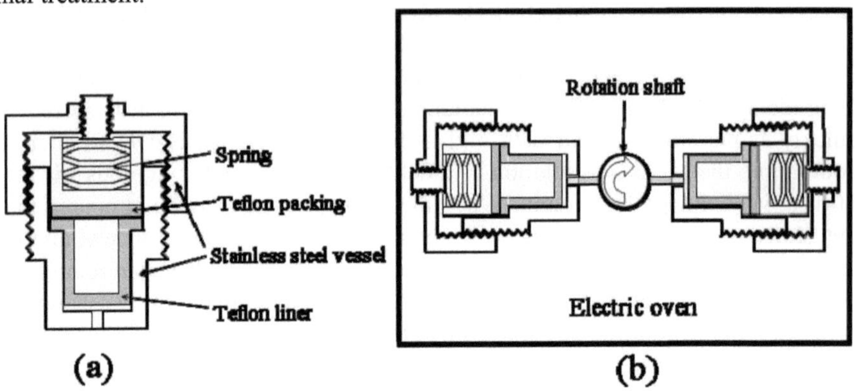

Fig. 2. Schematic diagram of hydrothermal treatment system. (a) Teflon lined and sealed autoclave; (b) rotation shaft for stirring set in an electric oven.

An amount of 1 g of the poorly crystalline Na-birnessite and 15 mL of 1 M NaOH solution are placed in the autoclave, which is then sealed. The autoclave is fixed on the rotation shaft in the electric oven. To keep balance of the rotation shaft, another autoclave is fixed at the opposite side of the rotation shaft as shown in Fig. 2b. The temperature of the electric oven is kept at 150°C for 12 h. The autoclave is then cooled to room temperature, and the product is filtered, washed with distilled water, and freeze-dried for 24 h.

Comments

The by-product β-MnOOH can be transformed to birnessite by hydrothermal treatment. Therefore, highly pure birnessite is obtained after hydrothermal treatment.

Freeze-drying is recommended for the drying of the birnessite precipitate gel, because drying by normal methods results in aggregation of the crystals to large and hard particles. It is difficult to crush the hard aggregate particles.

Characterization

The structure and purity of the birnessite is investigated by powder X-ray diffraction analysis. The major diffraction peaks of Na-birnessites are shown in Table 1. The poorly crystalline Na-birnessite has a layered structure with a basal spacing of 1.00 nm before drying (wet sample). After drying at room temperature, the basal spacing decreases to 0.72 nm. The impurity of β-MnOOH shows a major peak at a d-value of 0.46 nm. The highly crystalline Na-birnessite has the same basal spacing (0.72 nm) as the dried poorly crystalline sample, but shows more intense diffraction peaks than the poorly crystalline birnessite.

Table 1. The d-values of major XRD diffraction peaks of Na-birnessites

Product	d-values of diffraction peaks (nm)			
Poorly crystalline birnessite (wet)	1.004	0.503	0.336	0.248
Poorly crystalline birnessite (dried)	0.724	0.358	0.249	0.241
Highly crystalline birnessite	0.724	0.358	0.249	0.241

The crystal morphology is characterized by *scanning electron microscopy* (SEM). The highly crystalline Na-birnessite has a flower-like platy morphology with crystal sizes of about 5 μm in diameter and 0.2 μm in thickness, as shown in Fig. 3.

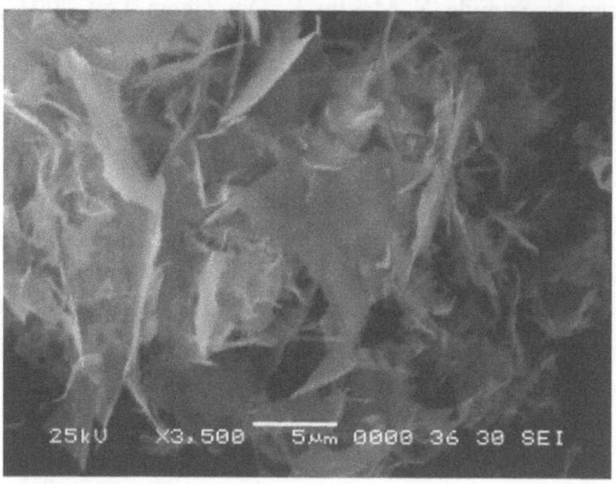

Fig. 3. SEM image of highly crystalline Na-birnessite.

The sodium and manganese contents in the products are determined by *atomic absorption spectrometry* after dissolving the product in a mixed solution of H_2SO_4 and H_2O_2, and the H_2O content by *thermogravimetric* and *differential thermal analysis* (TG-DTA). The interlayer water molecules are removed at about 150°C. The chemical composition is $Na_{0.31}MnO_{1.91}\cdot 0.7H_2O$ for the poorly crystalline birnessite, and $Na_{0.38}MnO_{1.95}\cdot 0.7H_2O$ for the highly crystalline birnessite, respectively.

References

[1] Q. Feng, H. Kanoh, K. Ooi, J. Mater. Chem. 1999, **9**, 319.
[2] S. Turner, P. R. Buseck, Science 1981, **212**, 1024.
[3] R. G. Burns, V. M. Burns, in Proc. Int. Symp. on Manganese Dioxide, Vol. 2, B. Schumm, H. M. Joseph, A. Kozawa, Eds, The Electrochemical Society, Cleveland, 1980, Chapter 6.
[4] R. Chen, P. Zavalij, M. S. Whittingham, Chem. Mater. 1996, **8**, 1275.
[5] Q. Feng, H. Kanoh, Y. Miyai, K. Ooi, Chem. Mater. 1995, **7**, 1226.
[6] M. Tsuji, S. Komarneni, Y. Tamaura, M. Abe, Mater. Res. Bull. 1992, **27**, 741.
[7] M. T. Sabine, A. W. Hewat, J. Nucl. Mater. 1982, **110**, 173.
[8] S. C. Wu, C. Chu, Atmos. Environ. 1972, **6**, 309.
[9] A. L. Cabrear, M. B. Maple, Appl. Catal. 1990, **64**, 309.
[10] M. Nitta, Appl. Catal. 1984, **9**, 151.
[11] S. Bach, J. P. Pereira-Ramos, N. Baffier, Electrochim. Acta 1993, **38**, 1695.
[12] R. Chen, P. Zavalij, M. S. Whittingham, Solid State Ionics 1996, **86-88**, 1.
[13] B. J. Aronson, A. K. Kinser, S. Passerini, W. H. Smyrl, A. Stein, Chem. Mater. 1999, **11**, 949.
[14] Q. Feng, K. Yanagisawa, N. Yamasaki, J. Porous Mater. 1998, **5**, 153.
[15] T. Rziha, H. Gies, J. Rius, Eur. J. Meneral. 1996, **8**, 675.
[16] Y.-F. Shen, R. P. Zerger, S. L. Suib, L. McCurdy, D. L. Potter, C.-L. O'Young, Science 1993, **260**, 511.
[17] D. C. Golden, C. C. Chen, J. B. Dixon, Science 1986, **231**, 717.
[18] Z. Liu, K. Ooi, H. Kanoh, W. Tang, T. Tomida, Langmuir 2000, **16**, 4154.
[19] R. Ma, Y. Bando, T. Sasaki, J. Phys. Chem. B 2004, **108**, 2115.
[20] Z. Tian, Q. Feng, N. Sumida, Y. Makita, K. Ooi, Chem. Lett. 2004, **33**, 952.
[21] P. Strobel, C. Mouget, Mater, Res. Bull. 1993, **28**, 93.
[22] R. Giovanoli, P. Burki, M. Giuffredi, W. Stumm, Chimia 1975, **7**, 1226.
[23] J. Luo, S. L. Suib, J. Phys. Chem. B 1997, **101**, 10403.
[24] S. Hirano, R. Narita, S. Naka, Mater. Res. Bull. 1984, **19**, 1229.
[25] Y. Ma J. Luo, S. Suib, Chem. Mater. 1999, **11**, 1972.
[26] X. Yang, W. Tang, Q. Feng, K. Ooi, Crystal Growth & Design, 2003, **3**, 409.
[27] Q. Feng, E.-H. Sun, K. Yanagisawa, N. Yamasaki, J. Ceram. Soc. Jpn. 1997, **105**, 564.
[28] Q. Feng, L. Kiu, K. Yanagisawa, J. Mater. Sci. Lett. 2000, **19**, 1567.
[29] Q. Feng, Y. Higashimoto, K. Kajiyoshi, K. Yanagisawa, J. Mater. Sci. Lett. 2001, **20**, 269.

Templated Carbon from Pyrolysis of Pyrene in Pillared Clay Matrices

G. Sandí, K. A. Carrado and R. E. Winans

Abstract Unique carbonaceous materials suitable as anodes for lithium secondary batteries have been synthesized and characterized. The synthesis is based on the use of pillared clays and various organic precursors, including pyrene. Pyrolysis was performed at 500, 700 and 1000 °C followed by inorganic matrix removal. X-ray powder diffraction data show that the carbon samples have a highly disordered structure.

Classification

form:	amorphous carbons
function:	anode materials for rechargeable batteries
preparation:	intercalation in pillared clays, templating
composition:	carbon, with residual hydrogen

Introduction

Carbon is an excellent candidate for negative electrodes in lithium secondary batteries. This is because carbon can take the form of lithium intercalation compounds. Two important attributes of carbon compared to lithium metal are its stability to electrolyte decomposition and an increase in lithium diffusivity. The risk of dendrite penetration of the separators present in the lithium metal batteries is then eliminated.[1-5]

Extensive efforts have been dedicated to the research and development of different carbonaceous materials that are able to deliver high specific capacity (mAh/g), high cyclic efficiency and long cycling life.[6-13] The structure of the carbon is a major factor in the intercalation of lithium, both in how much can be intercalated and at what voltage. The electrochemical intercalation of lithium in graphitic carbon can be described by the following equation:

$$Li^+ + 6C + e^- \underset{oxidation}{\overset{reduction}{\Leftrightarrow}} LiC_6$$

One of our main objectives was to design carbon electrodes with predictable porosity and surface area characteristics. Inorganic templates were employed for the synthesis of the carbons from polymeric precursors. Tomita et al. [14] have reported the formation of carbon using inorganic templates, where polyacrylonitrile was carbonized at 700°C yielding thin films with relatively low surface area. Low surface areas are desirable to reduce the irreversible lithium storage in the first cycle.

The templates described in the present work are pillared clays (PILCs). These modified clays have inorganic supports between the silicate layers that help to prevent the collapse of the layers upon heat treatment. Pyrene, the organic precursor for these studies, is dispersed in benzene between the PILC layers and subsequently pyrolyzed. After elimination of the inorganic matrix via demineralization, the layered carbons display holes due to the pillaring Al_{13} cluster where lithium diffusion may be able to occur. In a previous study of these materials using small angle neutron scattering, Winans and Carrado [15] showed that the diameter of the holes was about 15 Å, which is the approximate size of the Al_{13} pillar. Lithium should be able to diffuse rapidly through such a molecularly porous carbon.

Materials

- Montmorillonite clay, Bentolite L, from Southern Clay Products, Inc., which was purified commercially to remove most of the Fe.
- Chlorhydrol® 50% (the precursor of $[Al_{13}O_4(OH)_{24}(H_2O)_{12}]^{7+}$) from Reheis Chemical Co., used as received.
- Ammonium hydroxide solution, ACS Reagent 28-30% from Aldrich, used as received.
- Pyrene powder 99.9 % purity from Aldrich, used as received.
- Benzene 99.5% purity from Aldrich, used as received.
- Hydrofluoric acid (conc.)

Safety and Disposal

Safety and handling instructions for the chemicals are found in the corresponding materials safety data sheets (MSDS). The loading of the clay with the pyrene solution should be performed in a hood because of the toxicity of benzene (used to solubilize pyrene). The reaction of the composite material with HF is extremely exothermic. When using HF, proper protection equipment should be used (face shield, goggles, and gloves).

Procedures

A. Preparation of Pillared Clay (PILC)

1. A 5% v/v solution of Chlorhydrol® was prepared with 20 mL of Chlorhydrol® 50% v/v, which was then charged into a 200 ml graduate cylinder. To this was added 180 ml of dionized water.
2. 20 grams of Bentolite L (Ca-Bentonite) was slurried into 2 L of dionized water and heated to 65-70°C with constant stirring. Mechanical stirring might be necessary.
3. The Chlorhydrol® 5% v/v solution is added once the clay slurry reaches between 65-70°C, which normally occurs about 60-90 min after the dilution.
4. After 15 min, approx. 8 mL of 0.2 M NH_4OH was added until the pH was about 5.5 (checked with microfine paper). The pH should remain at this value throughout.
5. The slurry was stirred at 65-70°C for 1 h, then removed from the heat source and the clay were allowed to settle without stirring for a couple of hours.
6. The slurry solution was then decanted.
7. The solids were rinsed twice with deionized water (approx. 500 mL each). The final decant should be nearly clear.
8. The solids were air-dried. They may be dried in a large watch glass on top of a hot oven, but should not actually be placed directly in the oven. Temperature must be around 50°C and should not exceed this.
9. The dried solids were ground in a mortar. The X-ray powder diffraction (XRD) d-spacing of the 001 peak should be about 19.5 Å.
10. The solids were calcined at 400°C in flowing air for 4 h. The XRD d-spacing should be about 18.6 Å. This calcination step is what dehydroxylates the aluminum-containing pillaring species into more stable alumina pillars:

$$n[Al_{13}O_4(OH)_{24+X}(H_2O)_{12-X}]^{(7-x)+} \Rightarrow 6.5n\,Al_2O_3 + (7-x)H^+$$

Characterization

Specific surface area 216±10 m^2g^{-1} (determined by N_2 sorption and five-point BET analysis). The sample was evacuated at 80 mTorr overnight at 120°C.

XRD d_{001} spacings are (Fig. 1): Bentolite L, 14.9 Å; PILC not calcined, 19.5 Å, PILC calcined, 18.6 Å.

Comments

In step 3, the solution of Chlorhydrol® 5% v/v must be added to the clay slurry 60-90 min after dilution. Aging is important when pillaring clays.

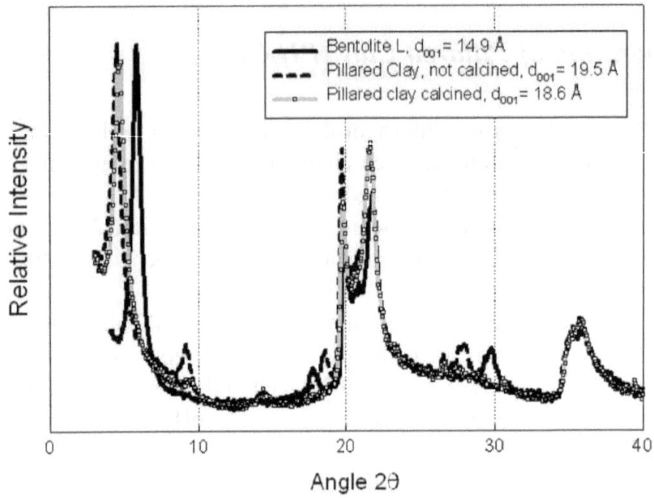

Fig. 1. XRD of Bentolite L, PILC not calcined, and PILC calcined, as indicated in the inset.

Most montmorillonites around the world are fairly similar and should, if purified, produce fairly similar results. The samples vary in terms of cation exchange capacity, but most are around 80 meq/100 g. This value is important in terms of the final pillar density, which therefore affects the amount of loaded carbon to a small degree. Another variable factor is the exchangeable cation itself. The montmorillonite used in this study has Ca^{2+} ions as the predominant interlayer cation. The pillaring efficiency can be affected by this cation, which is Na^{+} in many pillared clays. In addition, the surface area of the clay used, might affect the results. Purified and sieved to less than 2 μm, the surface area ordinarily is 80 m^2g^{-1}, which corresponds to the conditions used for the reported experiments.

In theory, the procedure can be transferred to all types of pillared clays, however, the yield and surface areas of the carbon may vary somewhat based on the type used.

B. Loading of the Pillared Clay with Organic Compounds

1. A 0.10 M solution of pyrene was prepared by dissolving 4.04 g of pyrene powder in 200 mL of benzene in a 500 mL round bottom flask. This was stirred with a magnetic bar.
2. Once the pyrene had totally dissolved, 2 g of the pillared clay obtained in part A was slowly added with constant stirring at room temperature overnight.
3. The following day, the stirring was stopped, the contents were allowed to settle, and the solution was decanted and was rinsed twice with 50 mL of ben-

zene to remove excess pyrene. A rotary evaporator may be used to remove all remaining benzene.

Comments

The pyrene-loaded powder sample (step 3), should have a green color after drying due to the electron transfer in the aromatic rings.

C. Pyrolysis, and Removal of the Inorganic Matrix

1. Approximately 2 g of the pyrene-loaded PILC was heated at 700°C under nitrogen for about 3 h. The sample turned black as an indication of the carbon formation.
2. The composite sample in was dissolved 100 mL of previously cooled concentrated HF with continuous stirring for 1 h.
3. The solution was centrifuged using Teflon tubes, and the sample washed with distilled water until the pH was ≥ 2.5.
4. The sample from step 3 was refluxed using 100 mL of concentrated HCl for 2 h.
5. The resulting carbon sample was centrifuged and washed until the pH was at least 5.5.
6. The carbon sample was dried in a conventional oven at 100 °C overnight.

Characterization

Elemental analysis: C, 94.1; H, 3.8; N, 2.1 (for C,H,N determination the sample was dried at 60 °C in vacuo prior to analysis).

Specific surface area 11 ± 5m^2g^{-1} (determined by N_2 sorption and five-point BET analysis). Depending on the HF used – concentrated or diluted – the surface area is variable over a wider range. The sample was evacuated at 80 mTorr overnight at 120 °C.

XRD: the d_{002} spacing of this carbon is 3.40 Å (Fig. 2).

Comments

The concentrated HF should be cooled in an ice bath prior to adding the composite material. Add the material very slowly because the reaction is extremely exothermic. The final specific surface area seems to be dependent on the concentration of HF used during dissolution of the composite sample.

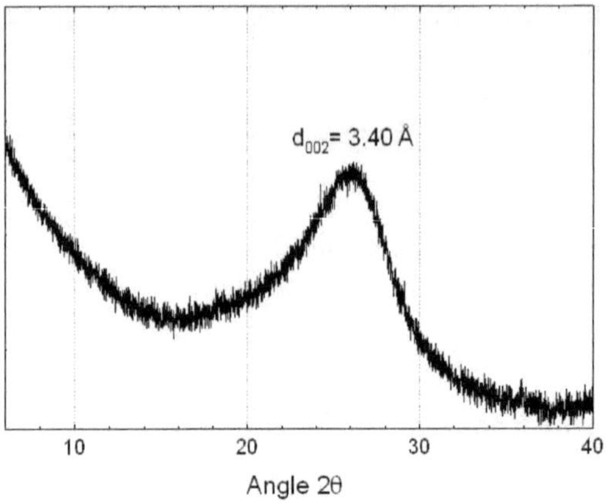

Fig. 2. XRD of carbon sample derived from pyrene-PILC.

References

[1] D. Aurbach, I. Weissman, A. Zaban, O. Chusid, Electrochim. Acta. 1994, **39**, 51.

[2] J. R. Dahn, A. K. Sleigh, H. Shi, J. N. Reimers, Q. Zhong, B. M. Way, Electrochim. Acta. 1993, **38**, 1179.

[3] R. Fong, U. von Sacken, J. R. Dahn, J. Electrochem. Soc. 1990, **137**, 2009.

[4] T. Iijima, K. Suzuki, K. Matsuda, Denki Kagazu. 1993, **61**, 1383.

[5] Y. Matsuda, J. Power Sources 1993, **43-44**, 1.

[6] J. R. Dahn, A. K. Sligh, H. Shi, B. W. Way, W. J. Weydanz, J. N. Reimers, Q. Zhong, U. von Sacken, in Lithium Batteries-New Materials, Developments and Perspectives, G. Pistoia, Ed., Elsevier, Amsterdam 1994, p. 1.

[7] N. Imanishi, H. Kashiwagi, T. Ichikawa, Y. Takeda, O. Yamamoto, M. Inagaki, J. Electrochem. Soc. 1993, **140**, 315.

[8] M. Mohri, N. Yanagisawa, Y. Tajima, H. Tanaka, T. Mitate, S. Nakajima, M. Yoshida, Y. Yoshimoto, T. Suzuki, H. Wada, J. Power Sources 1989, 26, 545.

[9] T. Ohzuku, Y. Iwakoshi, K. Sawai, J. Electrochem. Soc. 1990, **140**, 2490.

[10] K. Tatsumi, N. Iwashita, H. Sakaebe, H. Shioyama, S. Higuchi, J. Electrochem. Soc. 1995, **143**, 716.

[11] M. Wilson, J. R. Dahn, J. Electrochem. Soc. 1995, **142**, 326.

[12] K. Sleigh, U. von Sacken, Solid State Ionics 1992, **57**, 99.

[13] R. Kanno, Y. Takeda, T. Ichikawa, K. Nakanishi, O. Yamamoto, J. Power Sources 1989, **26**, 535.

[14] T. Kyotani, N. Sonobe, A. Tomita, Nature 1988, **331**, 331.

[15] R. E. Winans, K. A. Carrado, J. Power Sources 1995, **54**, 11.

Fiberous Carbon from Sepiolite Clay and Propylene

G. Sandí, K. A. Carrado and R. E. Winans

Abstract A pure carbon-based material with applications in electrochemical processes was synthesized using sepiolite clay as a template and propylene as the organic precursor. Carbon fibers (1-1.5 μm long) were obtained whose orientation and shape resemble that of the original clay. The results indicated that there is a correlation of the reversible specific capacity obtained and the surface properties of the template sepiolite. The electrochemical performance for anode Li cells is related to the surface chemical properties rather than the BET surface area.

Classification

form:	amorphous carbons
function:	anodes for rechargeable batteries
preparation:	gas phase
composition:	carbon, with residual hydrogen

Introduction

Carbon represents a very attractive material for electrochemical applications, especially for the storage of energy due to different allotropes (graphite, diamond, fullerenes/nanotubes), various microtextures (more or less ordered) the degree of graphitization, a rich variety of dimensionality (from 0 to 3D), and the ability for existence in different forms (from powders to fibers, foams, fabrics, and composites).

The successful utilization of a carbon host to store lithium ions in the rechargeable negative electrode has lead to the commercial development of Li-ion cells. Storage of Li in carbon to form the negative electrode in Li-ion cells occurs by different mechanisms. Danh et al. [1] proposed three mechanisms for lithium insertion in carbonaceous materials. The physical mechanism for this insertion depends on the carbon type. Lithium intercalates in layered carbons such as graphite, and it adsorbs on the surfaces of single carbon layers in non-graphitizable hard carbons. Lithium also appears to reversibly bind near hydrogen atoms in carbonaceous materials containing substantial hydrogen, which are made by heating organic precursors to temperatures near 700°C. Each of these three classes of materials appears suitable for use in advanced lithium batteries. More recently, Sandí et al. proposed a mechanism based on the concept that carbons with curved lattices can exhibit enhanced lithium capacity over that of graphite. This idea was underscored by computational studies of endohedral lithium complexes of buckminster fullerene, C_{60}.[2,3] It was found that the interior of the C_{60} molecule was large enough to easily accommodate two or three lithium atoms. Furthermore, the curved ring structure of the C_{60} molecule facilitated the close approach, 2.96 Å, of the lithium atoms even in the trilithiated species. This is significantly closer than the interlithium distance in the stage-one graphite intercalation compound LiC_6 and suggests that lithium anode capacities may be improved over graphitic carbon by synthesizing carbons with curved lattices that approximate a portion of a buckey ball.[4]

Sepiolite is a phyllosilicate clay and contains continuous two-dimensional sheets of tetrahedral silicate. It differs from other clays in that it lacks continuous sheets of octahedral constituents. Instead, its structure can be considered to contain ribbons of 2:1 phyllosilicate structure, with each ribbon linked to the next by inversion of SiO_4 tetrahedra along a set of Si-O-Si bonds. In this framework, rectangular channels run parallel to the x-axis between opposing 2:1 ribbons, which results in a fibrous morphology with channels running parallel to the fiber length. Channel diameters are 3.7 x 10.6 Å in sepiolite. Individual fibers generally range from about 100 Å to 4-5 μm in length, 100-300 Å width, and 50-100 Å thickness. Inside the channels are protons, coordinated water, a small number of exchangeable cations, and zeolitic water.

Carbonaceous materials with enhanced lithium capacity have been derived from propylene upon incorporation in the vapor phase in the channels of sepiolite, taking advantage of the Brønsted acidity in the channels to polymerize olefins.[5] After removal of the clay, carbon fibers (1-1.5 μm long) are obtained whose orientation and shape resemble that of the original clay.

Aurbach et al. [6] and Fong et al. [7] suggested that low surface area carbons are favorable for practical applications, since the amount of lithium consumed in the formation of the passivating layer that contributed to the irreversible capacity was proportional to the surface area of the carbon. However, Nasrin et al. [8] showed that porous high surface area carbons proved to be excellent candidates for lithium ion batteries. A procedure to prepare well-defined fibrous carbons is presented.

Materials

- Sepiolite clay, from The Tolsa Group, Spain, used as received.
- Propylene gas 99.95% purity, from AGA

- Hydrofluoric acid, 73% m/m, ACS reagent from Aldrich
- Hydrochloric acid, 37% m/m, ACS reagent from Aldrich
- Quartz boats
- Nitrogen gas 99% purity, from AGA

Safety and Disposal

Safety and handling instructions for the chemicals are found in the corresponding materials safety data sheets (MSDS). The reaction of the composite material with HF is extremely exothermic. When using HF, proper protection equipment should be used (face shield, goggles, and gloves).

Procedures

A. Preparation of Propylene-loaded Sepiolite

1. Approximately 3 g of sepiolite clay is placed into quartz boats, and the quartz boats are placed inside a quartz tube, as shown in Fig. 1 (it is easier to manipulate two single boats than a long boat. Spills are thus avoided when getting the boats out of the quartz tube).
2. The quartz tube and sample are flushed with nitrogen for about 30 min to remove air.
3. The propylene tank is opened and the gas flowed at 5 $cm^3 min^{-1}$.
4. The sample is then heated. The temperature of the oven should be gradually increased from room temperature (at about 5°C min^{-1}) to 700 °C. The oven is then held at that target temperature for 4 h.

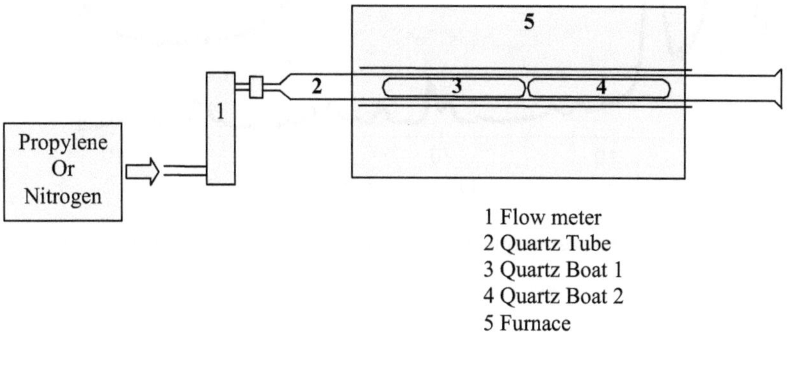

Fig. 1. Schematic representation of the system used to synthesize the carbons.

5. Close the propylene tank, open the nitrogen tank, and turn off the heater.
6. Let the composite sample cool to room temperature under a nitrogen flow.
7. Once the sample is at room temperature the nitrogen tank should be closed.

Characterization

Specific surface area of sepiolite clay: 216±10 m^2g^{-1} (determined by N_2 sorption and five-point BET analysis). The sample was evacuated at 80 mTorr overnight at 120 °C.

XRD pattern was determined using a Rigaku Miniflex, with Cu K_α radiation (wavelength 1.54051 Å) and a NaI detector at a scan rate of 0.5° 21/min. d_{110} spacing of sepiolite: 12.1 Å, d_{002} spacing of composite after loading with propylene: 3.7 Å, d_{002} of carbon sample 3.57 Å (Fig.2).

Comments

As evidenced by the disappearance of the 110 peak of sepiolite (see Fig. 2), there is marked loss of crystallinity upon heating the clay in the presence of propylene at 700 °C. The incorporation of carbonaceous precursor is indicated by a new broad peak at about 24° 2θ.

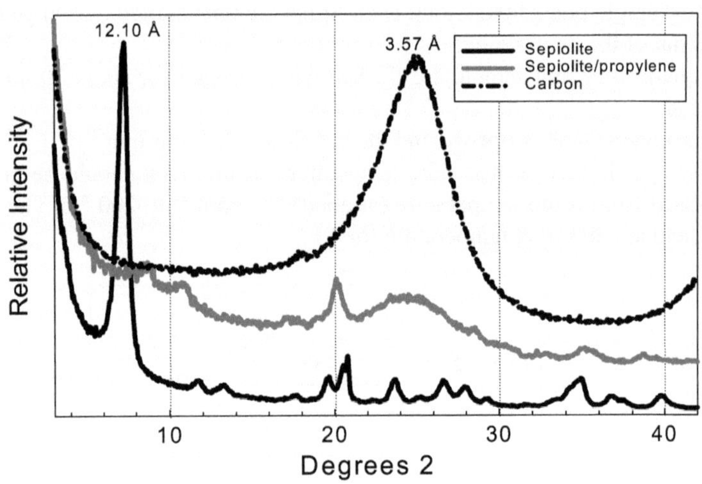

Fig. 2. XRD of sepiolite, composite sepiolite/carbon, and carbon, as indicated in the inset.

B. Pyrolysis and Dissolution of the Propylene-loaded Sepiolite

1. The composite sample is dissolved in 100 mL of previously cooled HF with continuous stirring for 1 h.
2. The solution is then centrifuged using Teflon tubes the sample washed with distilled water until the pH is at least 2.5.
3. Reflux the sample from step 2 using 100 mL of concentrated HCl for 2 h.
4. The resulting carbon sample should be centrifuged and washed until the pH is at least 5.5.
5. The carbon sample should be dried in a conventional oven at 100°C overnight.

Characterization

Elemental analysis: C, 93.9; H, 4.0; N, 2.1 (the sample is dried at 60°C in vacuo prior to analysis).

Specific surface area: 196±5 m^2g^{-1}. The sample was evacuated at 80 mTorr overnight at 120 °C.

XRD: d_{002} spacing of carbon: 3.57 Å (Fig. 2).

TEM (JEOL 100CXII Transmission Electron Microscope operating at 100kV): The selected area electron diffraction (SAED) pattern of the carbon fibers shows diffuse rings typical of amorphous carbon; no diffraction spots were observed (Fig. 3).

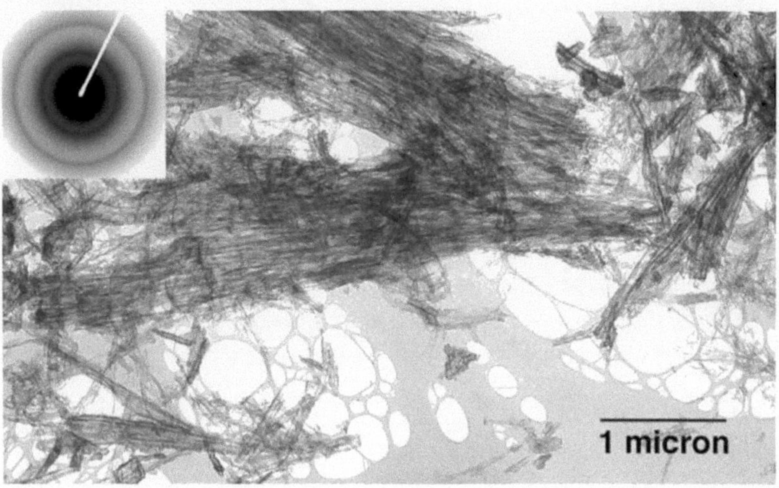

Fig 3. TEM of a carbon sample derived from sepiolite raw material after the clay has been removed.

Comments

The HF should be cool using an ice bath prior to adding the composite material. Add the material very slowly because the reaction is extremely exothermic. Upon removal of the inorganic matrix, there is only a broad peak at 3.57 Å (Fig. 2), which corresponds to the 002 reflection of disordered graphite.

References

[1] J. R. Dahn, T. Zheng, Y. H. Liu, J. S. Xue, Science 1995, **270**, 590.

[2] G. Sandí, R. E. Gerald II, L. G. Scanlon, K. A. Carrado, R. E. Winans, Mat. Res. Soc. Symp. Proc. 1998, **95**, 496.

[3] L. G. Scanlon, G. Sandí, Proc. 38th Power Sources Conf. 1998, 382.

[4] L. G. Scanlon, G. Sandí, J. Power Sources 1999, **81-82**, 176.

[5] G. Sandí, K. A. Carrado, R. E. Winans, C. E. Johnson, R. Csencsits, J. Electrochem. Soc. 1999, **146**, 3644.

[6] D. Aurbach, M. L. Daroux, P. W. Faguy, E. B. Yeager, J. Electrochem. Soc. 1987, **134**, 1611.

[7] R. Fong, U. von Sacken, J. R. Dahn, J. Electrochem. Soc. 1990, **137**, 2009.

[8] N. R. Khalili, M. Campbell, G. Sandí, W. Lu, I. V. Barsokov, J. New Mat. Electrochem. Sys. 2001, **4**, 267.

Aerosol Spray Synthesis of Porous Molybdenum Sulfide Powder

S. E. Skrabalak and K. Suslick

Abstract Highly porous, nanostructured MoS_2 spheres have been prepared by ultrasonic spray pyrolysis (USP), an aerosol synthesis technique.[1] An aqueous solution of ammonium tetrathiomolybdate, $(NH_4)_2MoS_4$, and colloidal silica, SiO_2, was ultrasonically nebulized using a household humidifier; the resulting aerosol droplets are heated in a furnace where solvent evaporation and precursor decomposition occurs, yielding a MoS_2/SiO_2 composite. Leaching of the colloidal SiO_2 with hydrofluoric acid, HF, results in a porous, high surface area MoS_2 network. The resulting material is a highly active hydrodesulfurization catalyst.

Classification

form:	porous, fine powder
function:	hydrodesulfurization catalyst
preparation:	aerosol spray, pyrolysis
composition:	MoS_2

Introduction

Aerosol syntheses and processing are common in materials science because of their ability to be scaled-up for industrial applications.[2] Ultrasonic spray pyrolysis (USP) is an attractive aerosol technique because it provides researchers with control over a wide range of experimental conditions. Fig. 1 shows a typical, laboratory-scale USP apparatus that is inexpensive and easily assembled. The process involves the atomization of one or more precursor solutions. Aerosol droplets are then carried by a gas (inert or reactive) into a furnace where solvent evaporation and precursor decomposition occurs. The product is collected in bubblers, and byproducts either remain dissolved in the collection solvent or are flushed out of the system by the carrier gas.

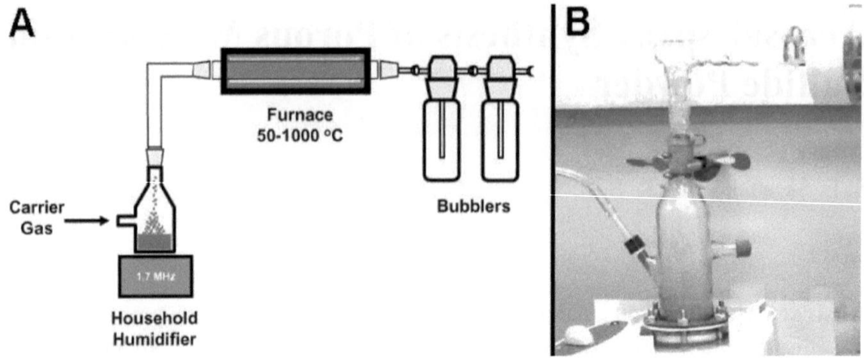

Fig. 1. (a) Typical laboratory-scale USP apparatus. (b) Photograph of atomization cell and base.

The droplets are individual micro-reactors: the size and morphology of the resulting material can be controlled by the size of and chemistry within the droplets. Eq. 1 shows the parameters that dictate droplet size, D_d. By changing the frequency of atomization or solvent/solution properties, droplet size can be controlled. Additionally, the final particle size, D_p, can be tuned by changing the concentration or the composition of the precursor solution, as indicated by Eq. 2. Typically, micron-sized particles are obtained; however, by adding surfactants or template material, particle morphology and porosity can be greatly altered. Here, two syntheses are presented. The first synthesis presents methodology for the production of micron-sized MoS_2 spheres of relatively low surface area (~40 $m^2\ g^{-1}$) through the thermal decomposition of $(NH_4)_2MoS_4$. The second synthesis incorporates a template, SiO_2, which upon chemical removal, yields high surface area (100-250 m^2g^{-1}), porous MoS_2. Both powders contrast greatly with MoS_2 synthesized by conventional techniques (Fig. 2A).

$$D_d = 0.34\left(\frac{8\pi\sigma}{\rho f^2}\right)^{1/3} \qquad (1)$$

D_d = *droplet diameter*
σ = *surface tension*
ρ = *density*
f = *atomization frequency*

$$D_p = \left(\frac{M D_d^{\,3} C_s}{1000\rho}\right)^{1/3} \qquad (2)$$

D_p = *particle diameter*
M = *molecular weight*
C_s = *molar concentration*
D_d = *droplet diameter*
ρ = *density*

The production of high surface area, porous MoS_2 is important because of its use as the standard hydrodesulfurization catalyst in the petroleum industry.[3] MoS_2 has a layered structure with a repeating motif of S-Mo-S sandwiched layers. Hydrodesulfurization, however, only occurs at the exposed edges of these layers.[4] Thus, the desulfurization activity of MoS_2 is greatly affected by the synthetic technique employed in its production. By using USP to prepare MoS_2, the catalytically active edge sites are increased substantially, improving the desulfurization properties.

Materials

- Ammonium tetrathiomolybdate $(NH_4)_2MoS_4$ from Aldrich (99.97%); used as received.
- SNOWTEX® ZL colloidal silica (~80 nm) 40 wt% solution in water; used as received.
- 49% HF diluted to 10% with ethanol (reagent grade).
- Deionized water.
- Ar gas (purity $\geq$ 99.99%).

Safety and Disposal

Safety and handling instructions for the chemicals are found in the corresponding materials safety data sheets (MSDS). Pyrolysis of $(NH_4)_2MoS_4$ generates hydrogen sulfide, H_2S; a bubbler of bleach or concentrated sodium hydroxide (aq) should be attached to the end of the product collection bubblers. The experiments should be conducted in a well-ventilated fume hood. Extra caution should be employed while handling HF to avoid contact with skin.

Procedures

The USP apparatus shown in Fig. 1A consisted of a household *Sunbeam*™ Ultrasonic humidifier base (model #696, 1.7 MHz) filled with water as a coupling medium. The base of the atomization cell has a 52 mm diameter opening with O-ring groove (ChemGlass #CG-138-02), over which a polyethylene membrane (2 mils, cut from a zip-lock plastic bag) was clamped between a Teflon donut and the O-ring (Fig 1B). The atomization cell had both a gas inlet and solution-addition arm. The atomization cell was placed directly above the humidifier transducer, and the glassware assembled as in Fig. 1. The flow tube through the furnace was 25 mm quartz, which permits the use of high furnace temperatures. A single zone furnace with a total heated region of ~30 cm was employed. Four collection bubblers (each 200 mL volume) were used in series and each was filled ~25% with deionized water. A final scrubber bubbler filled with either bleach or aqueous sodium hydroxide was placed at the end of the bubbler series.

A. Preparation of Low-porosity MoS_2

To 25 ml of deionized water, 0.25 g $(NH_4)_2MoS_2$ was added and dissolved. The solution was sparged with Ar for 30 min and then added to the atomization cell through the solution-addition arm. The USP system was flushed with Ar prior to solution addition and was continuously flushed with Ar while the furnace was heated to 700 °C. Once temperature was reached, the Ar flow was set to 940 ml min^{-1} (monitored with a calibrated rotometer) and the humidifier was turned on. A fine orange-red mist is observed and black powder deposits on the cooler furnace tube outlet as well as in the collection bubblers.

The contents of the collection bubblers were combined into tubes and centrifuged. The supernatant was decanted off the powder. The powder was then re-suspended in ethanol and transferred to a round bottom flask. The ethanol was removed by rotary evaporation, and the powder collected.

Characterization

Elemental analysis: Samples were vacuum dried at 110 °C for 24 h prior to analysis. The total weight percent of carbon, hydrogen, and nitrogen contaminants together was below 1%. Oxygen was below 2% (calculated by difference). The molybdenum to sulfur mole ratio was 2.0±0.1.

XRD: Four broad peaks were observed with d-spacings 6.35, 2.70, 2.54, and 1.57 Å which correspond to the [002], [100], [103], and [110] reflections of poorly crystalline hexagonal MoS_2. Spectra were obtained with a Rigaku D-MAX diffractometer using Cu Kα; crystallite sizes were determined using the Jade X-ray analysis software package. C-stacking height, calculated with the Scherrer equation using the [002] reflection, was ~36 Å.

BET surface area: N_2 isotherms (at 77 K) and surface area measurements were obtained with Quantachrome Instruments Nova 2200e Surface Area and Pore Analyzer. Three point analysis gave a surface area of 20–40 m^2g^{-1}.

SEM (Fig. 2): Hitachi S-4700 SEM operating at 10 kV.

B. Preparation of Porous MoS_2

To 25 ml of deionized water, 0.25 g $(NH_4)_2MoS_4$ and 1.20 g of SNOWTEX-ZL was added. The remainder of the synthesis and the powder collection was followed as described in part A.

Composite Etching. The collected powder was transferred to a plastic, HF resistant, centrifuge tube. Ca. 30 ml of 10% HF was added to the tube, the tube was sealed and spun for 24 h. The HF was then decanted off after centrifugation; the powder was re-suspended in water, centrifuged, and supernatant was removed.

This procedure was repeated until a neutral pH was obtained (~5 times). The powder was re-suspended in ethanol and transferred to a round bottom flask; the ethanol was removed by rotary evaporation, and the powder collected.

Characterization

Elemental analysis, XRD and SEM (Fig. 3 A and B): Same as in part A.

BET surface area: Same as in part A. Three point analysis gave a surface area of 100 m^2g^{-1}.

TEM (Fig. 3 C): Philips CM-12 TEM operating at 120 kV.

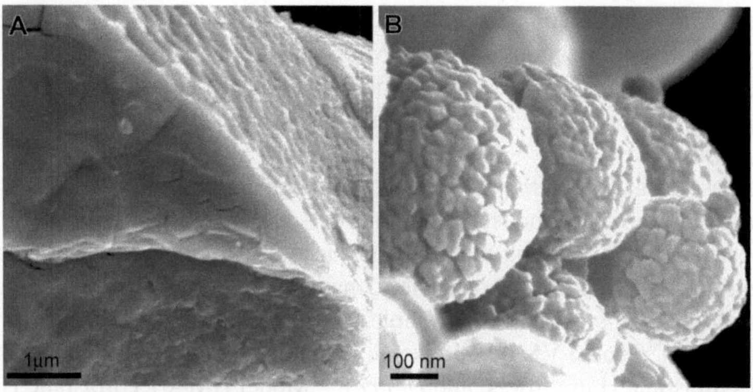

Fig. 2. (a) SEM image of conventional MoS_2. (b) SEM image of USP generated MoS_2 spheres.

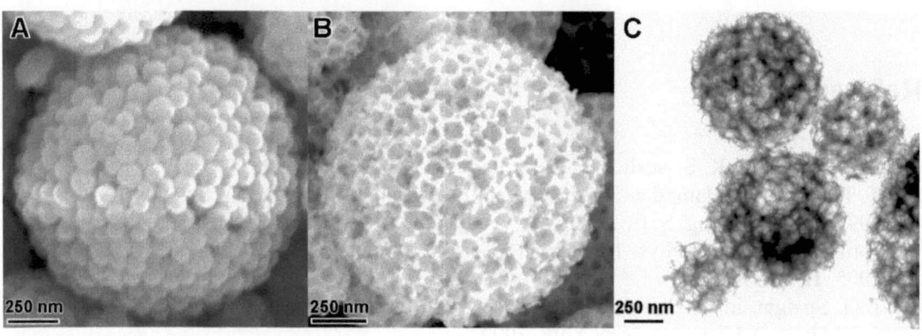

Fig. 3. (a) SEM image of MoS_2/SiO_2 composite made by USP. (b) SEM image of USP product after leaching of template SiO_2 with HF acid. (c) TEM image of USP product after leaching of template SiO_2 with HF acid.

Comments

(1) If poor atomization of solution is observed, it is likely due to air bubbles in the coupling water that are trapped under the polyethylene membrane. A syringe may be used to remove these.

(2) If a quartz furnace tube is unavailable, the furnace temperature should not exceed 450 °C with a Pyrex tube. The MoS_2 can be prepared even at 400°C; the XRD powder pattern of such material, however, may show peaks of elemental sulfur. Such sulfur can be removed by subsequent heating under an inert atmosphere.

(3) Custom bubblers, composed of ball and socket joints, were used in our apparatus. If standard bubblers are employed and connected via Tygon tubing, the weight percent carbon in the final samples was found to increase slightly.

(4) Colloidal silica is readily available in many sizes. This procedure can be modified by changing the size of the colloidal silica being used. This provides the researcher with the ability to modify the porosity and morphology of the MoS_2 to meet their specific application. MoS_2 surface areas of 250 m^2g^{-1} can be achieved. For different sized colloidal silica, the ratio of $(NH_4)_2MoS_4$ to silica may require optimization to prevent MoS_2 network collapse.

(5) The use of colloidal silica as a sacrificial template in aerosol syntheses can be extended to many other materials, resulting in unique morphologies and pore structures.[5]

(6) The base of a *Sunbeam*™ Model 696 Ultrasonic Humidifier was used in this study; however, other household humidifiers can be used without changing the properties of the resulting material significantly, so long as the ultrasonic frequency remains ~2 MHz. Fortunately, many commercially available ultrasonic humidifiers are similar. For example, successful results have been obtained with *Vicks*™ Model V5100 Ultrasonic Humidifier, *Sunbeam*™ Model 696 and 701 Ultrasonic Humidifiers, and *Holmes*™ Model HM461 Ultrasonic Humidifier.

References

[1] S. E. Skrabalak, K. S. Suslick, J. Am. Chem. Soc. 2005, **127**, 9990.

[2] T. T. Kodas, M. Hampden-Smith, Aerosol Processing of Materials, Wiley-VCH, New York, 1999. G. L. Messing, S. C. Zhang, G. V. Jayanthi, J. Am. Ceram. Soc. 1993, **76**, 2707. P. S. Patil, Mater. Chem. Phys. 1999, **59**, 185. Y. T. Didenko, K. S. Suslick, J. Am. Chem. Soc. 2005, **127**, 9990.

[3] J. G. Speight, in H. Heinemann (ed.) The Desulfurization of Heavy Oils and Residua, 2nd ed., Marcel Dekker, Inc., New York, 2000.

[4] S. J. Tauster, T. A. Pecoraro, R. R. Chianelli, J. Catal. 1980, **63**, 515. N. R. Dhas, A. Ekhtiarzadeh, K. S. Suslick, J. Am. Chem. Soc. 2001, **123**, 8310. P. Afanasiev, G. F. Xia, B. Gilles, B. Jouguet, M. Lacroix, Chem. Mater. 1999, **11**, 3216. N. Berntsen, T. Gutjahr, L. Loeffler, J. R. Gomm, R. Seshadri, W. Tremel, Chem. Mater. 2003, **15**, 4498.

[5] W. H. Suh, K. S. Suslick, J. Am. Chem. Soc. 2005, **127**, 12007.

Sonochemically Prepared Molybdenum Sulfide

A. Ryder and K. Suslick

Abstract A sonochemical preparation of high surface area, sub-micron molybdenum sulfide is described.[1] A slurry of elemental sulfur and molybdenum hexacarbonyl in isodurene was irradiated with high-intensity ultrasound (20 kHz, ~ 80 W/cm^2) under Ar followed by heating under vacuum to remove residual solvent and any unreacted precursor.

Classification

form:	amorphous, sub-micron powder
function:	hydrodesulfurization catalyst, lubricant
preparation:	sonochemistry
composition:	MoS_2

Introduction

Molybdenum sulfide is a layered material and the standard catalyst for industrial hydrodesulfurization (HDS) of petroleum and other fuel stocks.[2] The layers are repeating S-Mo-S sandwiches with only weak interactions between the sulfur-sulfur interfaces; this provides MoS_2 with its excellent lubrication properties. HDS catalysis by MoS_2, however, originates only from the exposed Mo atoms on the edges of the layers. A logical strategy to improve its catalytic activity would thus be to increase the relative amount of exposed edge sites. Conventional MoS_2 powder is difficult to grind below ~50 μm in size; worse, its preferred mode of cleavage is to split along the weak sulfur-sulfur interlamellar contacts, which does *not* increase the exposed Mo edges. By making MoS_2 (i) nanostructured to increase surface area, and (ii) largely amorphous to decrease the length scale over which MoS_2 layers are ordered, this can be achieved. Sonochemically synthesized materials have been shown to exhibit both sub-micron and amorphous character;[3,4] it follows that developing a procedure for sonochemical synthesis of MoS_2 creates a facile route to enhancing the catalytic activity of MoS_2.[1]

Ultrasonic cavitation (the formation, growth, and implosive collapse of gas-filled cavities (bubbles) in a liquid) is ultimately responsible for the unique properties of materials made using high-intensity ultrasound. In the presence of ultrasound, bubbles expand and contract with the sound field. During the rarefaction phase, volatile species surrounding the bubble diffuse inside to compensate for an increase in volume and decrease in pressure. This is possible since the time scale of expansion is relatively long (μs). However, the time scale for rapid, final compression is much shorter (nanoseconds) such that the resulting decrease in volume cannot be compensated for by mass and energy diffusion out of the bubble. As a result, extreme temperatures and pressures are achieved within the bubble (on the order of 5000 K and 300 bar),[5,6] as well as cooling rates well above 10^{10} Ks^{-1}. These unique and severe conditions can be used to drive chemical reactions within the bubble; one result is the synthesis of nanostructured, amorphous materials.

Metal carbonyls have high vapor pressures and low bond strengths, which make them an excellent choice as precursors for sonochemical reactions. Low vapor pressure solvents are also appropriate choices, as the solvent content within the bubble is minimized. This avoids extensive cushioning of bubble collapse, in turn resulting in higher achievable temperatures, pressures, and quenching rates.

Materials

- Sulfur (99%) purchased from Strem, used as received.
- $Mo(CO)_6$ (98%) purchased from Strem, used as received.
- Isodurene purchased from Aldrich, dried and distilled over Na, and degassed.
- Pentane purchased from Aldrich, dried and distilled over Na/benzophenone, and degassed.
- Ar, purified through molecular sieve and charcoal traps; the tank should have a stainless steel tubing connection to the regulator for connection to the reaction cell.
- VCX750 power supply (750W) and CV33 transducer purchased from Sonics & Materials (other sources: www.sigmaaldrich.com, www.aceglass.com, www.coleparmer.com).
- Ti horn with groove for o-ring, threaded end with removable tip, ½" diameter (Ace Glass product # 9814-25).
- Stainless steel tubing connected to a needle valve and ¼" Swagelok nut/Teflon® ferrule fitting for attachment to side arm on glass cell.
- Threaded Teflon® collar and o-ring for attachment of horn to glass cell.
- Glass cell with threads matching plastic collar and two ¼" OD side arms for gas inlet and outlet.
- 60 mL fine frit
- Glove box, < 0.5 ppm O_2

Safety and Disposal

Always handle $Mo(CO)_6$ in a fume hood. Never touch the ultrasonic horn when the ultrasonic processor is in use. Use a solid Ti horn (rather than one with a removable tip) when processing low- surface- tension liquids to avoid infiltration of solvent into the threaded portion of the tip; on occasion this can loosen the tip and cause damage to both horn and glassware.

Procedure

Preparation in Glove Box

In a glove box or other inert atmosphere apparatus, add 2.5 g of $Mo(CO)_6$, 0.75 g of sulfur, and 35 mL of isodurene to the glass cell used for sonochemical synthesis (Fig. 1). Slide the plastic collar and then the o-ring onto the Ti horn, and hand tighten the collar into the threading of the glass cell until it no longer turns (it should be air tight). To one side arm attach the valved tubing with Swagelok fitting, making sure the valve is closed. Seal the other sidearm such that air cannot penetrate the cell.

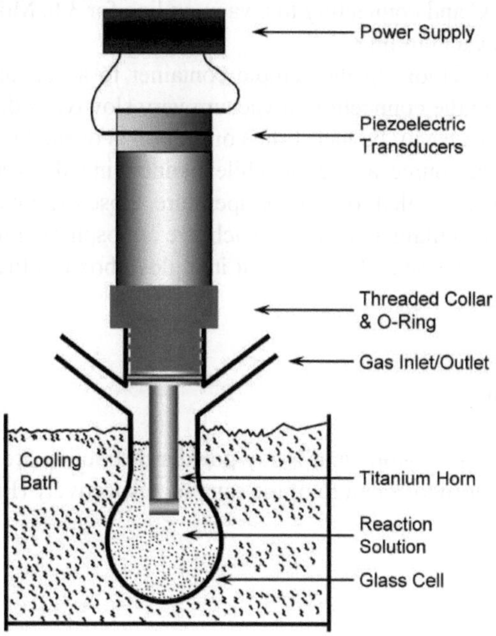

Fig. 1. Schematic of reaction cell and setup

Reaction in Fume Hood

Remove the cell from the glove box and connect the horn to the CV33 transducer. Connect the tubing on the side arm to a continuous flow of Ar, turn it on, and open the valve on the tubing connected to the glass cell side arm. Connect the gas outlet sidearm to a bubbler to monitor the Ar flow. Place the glass cell/horn/transducer assembly in a constant temperature bath (set to 60 °C) to a depth such that only the glass cell is immersed. Turn the power box on, making sure the transducer is connected to it; set the timer for 90 min. Slowly turn the power output up to ~80 W/cm^2 (the transducer can be calibrated calorimetrically to determine its actual power output at a given setting) and turn on the transducer. When 90 min have passed, seal the gas outlet and inlet.

Product Isolation

Product Recovery. Turn off the power supply and remove the horn/cell assembly from the transducer. Take this as well as filtering supplies and pentane (distilled and degassed) into the glove box. Filter the black slurry and wash it with six 30 mL aliquots of pentane. Transfer the black powder to a sealable container suitable for heating to 80 °C and connecting to a vacuum line for 3 h. Move the sample to a fume hood with a vacuum line.

Product Drying. Hook up the sample container to a vacuum line, and apply heat at 80 °C. Open the connection to vacuum very slowly, as the sample is a very fine powder and can easily be lost if the connection is opened too quickly. After 3 h, remove the heat source and cool while maintaining the vacuum connection. When the sample has cooled to room temperature, close the connection to vacuum and fill the sample container with Ar to achieve atmospheric pressure. Make sure the sample container is sealed and store it in a glove box or other appropriate vessel.

Characterization

Scanning electron microscopy and X-ray powder diffraction are used to verify the sub-micron and amorphous nature of the sample, respectively (Figs. 2 and 3).

Comments

(1) Chemicals may be purchased from a different supplier provided the purity is equivalent.

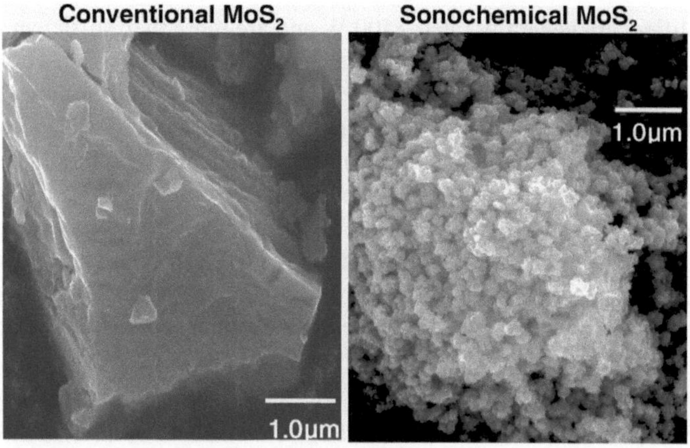

Fig. 2. SEM (Hitachi S800) of sonochemically prepared MoS_2 compared to conventional MoS_2.[1]

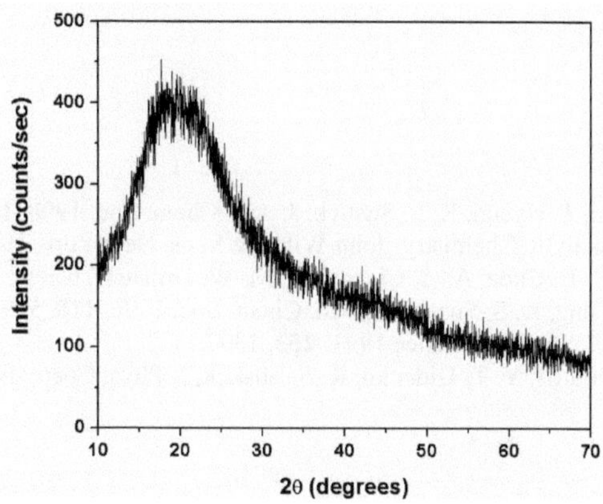

Fig. 3. XRD powder pattern of sonochemically prepared MoS_2 (Rigaku D-max diffractometer, Cu Kα radiation). The broad peak is from the glass slide on which the sample is mounted.

(2) The synthesis may be performed inside a glove box rather than a fume hood if the appropriate gas connections and cooling apparatus exist within the glove box.

(3) When tightening the Swagelok nut connection to the glass cell sidearm, wrenches should be used; hand tightening will result in an air leak. Care should be taken not to over tighten, however, as this will result in cracking the cell.

(4) After sonication for 90 min, the reaction yield is approximately 40%. While increasing sonication time will increase the overall yield, this time should not exceed 90 min as ultrasonic decomposition of the solvent will lead to increased carbon contamination.

(5) Nearly all commercially available ultrasonic horns (also known as "cell disruptors") function in the region of 20 kHz and have very similar designs; the

amplifiers are generally rated >500 W and are more than sufficient for generating 100 W/cm^2 with a 1 cm diameter horn. We generally find very similar results in general with units from Branson, Mysonix, and Sonics & Materials; the last of which is sold also by Aldrich Chemicals and by ACE Glass. In laboratory use, four important items must be remembered: First, the cavitation zone only extends ~3 cm from the horn surface, so a small horn in a large vessel is not effective; for larger scale reactions, flow reactors are commercially available or can be easily constructed. Second, too high an amplitude setting (i.e., "too much power") is counterproductive: at too high a setting, the horn becomes surrounded by a permanent cloud of large bubbles (mm diameter) and the horn becomes uncoupled from the liquid, i.e., the ultrasonic intensity getting into the liquid will actually diminish. Third, ultrasonic irradiation degasses liquids, so it is often beneficial to sparge the liquid with gas (usually Ar). Fourth, temperature control of the sonicated liquid is critical to avoid rapid overheating of the bulk liquid, which increases the solvent vapor pressure and suppresses the intense local heating within the collapsing bubble.

References

[1] M. M. Mdleleni, T. Hyeon, K. S. Suslick, J. Am. Chem. Soc. 1998, **120**, 6189.
[2] B. G. Gates, Catalytic Chemistry. John Wiley & Sons, New York, 1992.
[3] K. S. Suslick, S.-B. Choe, A. A. Cichowlas, M. W. Grinstaff, Nature 1991, **353**, 414.
[4] T. Hyeon, M. Fang, K. S. Suslick, J. Am. Chem. Soc. 1996, **118**, 5492.
[5] E. B. Flint, K. S. Suslick, Science 1991, **253**, 1397.
[6] W. B. McNamara III, Y. T. Didenko, K. S. Suslick, J. Phys. Chem. B 2003, **107**, 7303.

Doped Manganites

J. Spooren and R. I. Walton

Abstract Two doped, mixed-valent lanthanum manganese oxides, $La_{0.5}Ba_{0.5}MnO_3$ and $La_{0.5}Sr_{0.5}MnO_3$ were prepared using mild hydrothermal synthesis in one step at 240°C. The fine polycrystalline powders are phase-pure, and were indexed on a primitive cubic unit cell.

Classification

form:	polycrystalline powders
function:	ferromagnetic material, magnetoresistive material, redox catalyst
preparation:	hydrothermal synthesis
composition:	$La_{0.5}Ba_{0.5}MnO_3$, $La_{0.5}Sr_{0.5}MnO_3$

Introduction

Doped lanthanide manganites of the general formula $Ln_{1-x}A_xMnO_3$ (Ln = lanthanide ion, A = alkali earth metal ion) have attracted much interest from materials scientists, chemists and physicists owing to their wide-ranging magnetic and electronic properties. These properties are intimately linked to the crystal structures of the solids, which in turn depend upon the precise level of doping and the relative size of the metal ions. The materials all adopt the ABO_3, perovskite-type structure with varying degrees of distortion depending on a subtle balance of not only the size-match of the metal ions occupying the A and B sites, but also the relative sizes of A-site atoms in complex, multinary materials, and the amounts of Mn(III) and Mn(IV), the former being Jahn-Teller distorted. For example $LaMnO_3$ has orthorhombic symmetry and is an antiferromagnetic insulator, whereas $La_{0.5}Ba_{0.5}MnO_3$ is cubic and a ferromagnetic material at low temperatures. The 'half-doped' materials $Ln_{0.5}A_{0.5}MnO_3$ have attracted particular attention as they often exhibit the phenomenon of giant magnetoresistance, whereby the application of a magnetic field reduces the resistivity by several orders of magnitude.[1]

Members of this family of materials have also been used as redox catalysts,[2] and as electrode materials for solid-oxide fuel-cells.[3]

The usual method for the synthesis of the doped manganites is the conventional ceramic method of solid-state chemistry, whereby metal oxides or carbonates containing the desired metals are ground in stoichiometic proportion and fired at elevated temperature (> 1000°C). The reaction is deemed complete when powder X-ray diffraction data from the solid product show a phase-pure sample of the desired material. This procedure is lengthy and often requires repeated cycles of heating and regrinding to achieve sample homogeneity. In the case of the doped manganites, this is particularly true since at least three metal-oxide precursors are required. The ceramic method is also problematic when control of metal oxidation state is crucial to controlling the properties of the solid: annealing in controlled gas atmospheres is often necessary to stabilize a particular phase. This is encountered during the synthesis of the doped lanthanum manganites, since the Mn(III)/Mn(IV) ratio in the material is affected by non-stoichiometry: materials with composition $Ln_{1-x}A_xMnO_{3+\delta}$ are commonly formed.

The hydrothermal method, widely used for the synthesis of zeolites and other microporous materials, has considerable utility in the preparation of complex mixed-metal oxides. It has been shown that it is possible to prepare materials belonging to the $Ln_{1-x}A_xMnO_3$ family by a one-step hydrothermal procedure at 240°C.[4-8] The advantage of the hydrothermal method, aside from the low temperatures and short reaction times, is that fine powders are formed. This is important for many practical applications of the materials. In addition, the use of a comproportionation reaction in solution allows the desired manganese oxidation state to be dictated by choice of chemical reagents. Here we use the solution reaction between MnO_4^- and Mn^{2+} in a 3:7 ratio to give an average manganese oxidation state of 3.5, as desired in the doped manganites.

The hydrothermal reaction is performed in highly concentrated KOH as a mineraliser, allowing the rapid dissolution of the reagents.

Materials

- Potassium permanganate, purity 99%, purchased from Sigma-Aldrich and used as received.
- Manganese(II) sulfate monohydrate, purity > 98%, purchased from Sigma-Aldrich and used as received.
- Potassium hydroxide pellets, purchased from Sigma-Aldrich and used as received.
- Lanthanum nitrate hexahydrate, purity 99.999%, purchased from Sigma-Aldrich and used as received.
- Barium chloride dihydrate, purity > 99%, purchased from Sigma-Aldrich and used as received.

- Strontium sulfate (anhydrous), purity > 99%, purchased from Sigma-Aldrich and used as received.
- Distilled water
- 23 mL Parr-type, Teflon™-lined hydrothermal autoclave.
- Hydrochloric acid 37 wt % in water purchased from Sigma-Aldrich and diluted to ~ 2 M.
- Potassium iodide, purity > 99%, purchased from Sigma-Aldrich and used as received.
- 0.1 N sodium thiosulfate volumetric standard, purchased from Sigma-Aldrich and used as received.

Safety and Disposal

Safety and handling instructions for the chemicals are found in the corresponding materials safety data sheets (MSDS). Autoclaves should not be opened until they have been cooled to room temperature after a reaction has been performed.

Procedures

A. Preparation of $La_{0.5}Ba_{0.5}MnO_3$

Fresh solutions of 0.300 M $KMnO_4$, 0.350 M $MnSO_4$, 0.350 M $BaCl_2$, and 0.350 M $La(NO_3)_3$ were prepared. The actual level of hydration of the metal salts must be checked by gravimetric or thermogravimetric analysis to allow accurate preparation of these solutions. 2.00 mL of the $MnSO_4$ solution were mixed with 1.43 mL of the $BaCl_2$ solution and 1.43 mL of the $La(NO_3)_3$ solution in the Teflon™ liner of the autoclave with stirring. To this mixture was added 1.00 mL of the $KMnO_4$ solution, and finally ~7.00 g of KOH pellets with continued stirring. The final composition of the mixture was $7Mn^{2+}$: $3MnO_4^-$: $5Ba^{2+}$: $5La^{3+}$: 1250 KOH : 3256 H_2O. The complete mixture was stirred for 30 min before sealing the Teflon™ liner, containing the reagents, in the stainless-steel autoclave. The percentage fill of the autoclave is close to 50 %, and this level should be maintained if the reaction is scaled to larger reaction vessels. The autoclave was kept at 240°C for 24 h, followed by cooling at 2°C min^{-1} to room temperature. The solid product, a fine black powder, was isolated by suction filtration, washed with distilled water and acetone, and dried in air at 100°C for 24 h.

Characterization

Powder X-ray diffraction data (Fig. 1, Table 1) were indexed using a primitive cubic unit cell with $a = 3.9092(5)$ Å.

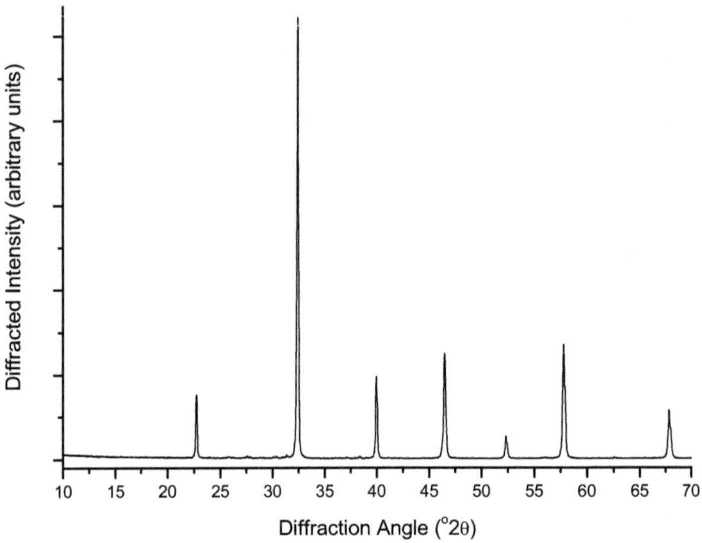

Fig. 1. Powder X-ray diffractogram from $La_{0.5}Ba_{0.5}MnO_3$ prepared by hydrothermal synthesis.

Table 1. Indexed powder X-ray diffraction data of $La_{0.5}Ba_{0.5}MnO_3$.

$2\theta_{obs}$*	I_{obs}	$2\theta_{calc}$	*hkl*
22.7485	13.9	22.7467	001
32.4009	100.0	32.3873	011
39.9623	19.7	39.9443	111
46.4892	33.5	46.4575	002
52.3648	6.8	52.3298	012
57.8262	37.9	57.7687	112
67.8039	19.4	67.8034	202

* Cu $K_{\alpha1}/K_{\alpha2}$ radiation, recorded from 5-70 °2θ.

Iodometric titration can be used to verify the average manganese oxidation state of a material of general composition $La^{III}_{1-x}A^{II}_{x}Mn^{III}_{1-\alpha}Mn^{IV}_{\alpha}O_{3\pm\delta}$ and confirm the oxygen stoichoimetry.[9] Around 0.1 g of $La_{0.5}Ba_{0.5}Mn_{1-\alpha}Mn_{\alpha}O_{3\pm\delta}$ is accurately weighed and dissolved in a mixture of 10 mL of a 10 wt% KI solution and 2.5 mL of 2 M HCl. The following reactions occur:

$$2\,Mn^{III} + 2\,Cl^- \rightarrow Cl_2 + 2Mn^{II}$$
$$Mn^{IV} + 2\,Cl^- \rightarrow Cl_2 + Mn^{II}$$
$$Cl_2 + 2\,I^- \rightarrow 2\,Cl^- + I_2$$

The iodine formed by these reactions is titrated against a 0.1 N sodium thiosulfate volumetric standard. Three drops of starch are added near the endpoint as an indicator. Since $3\pm\delta = (11+2\alpha)/4$, we can use the volume of added 0.1 N thiosulfate solution, V, to determine the amount of iodine released from m g of the manganite sample and hence the value of α. It can thus be shown for $La^{III}{}_{0.5}Ba^{II}{}_{0.5}Mn^{III}{}_{1-\alpha}Mn^{IV}{}_{\alpha}O_{3\pm\delta}$ that

$$\alpha = \frac{m - 27.31V}{0.8V - m}$$

For hydrothermal $La^{III}{}_{0.5}Ba^{II}{}_{0.5}Mn^{III}{}_{1-\alpha}Mn^{IV}{}_{\alpha}O_{3\pm\delta}$ values of $\alpha = 0.51(2)$ and $\delta = 0.008(13)$ were obtained, showing that the material has a composition close to the ideal stoichiometry $La_{0.5}Ba_{0.5}MnO_3$.

Thermogravimetric analysis recorded under an atmosphere of 10% H_2 in Ar showed a mass loss of 3.8 % between 100 and 600°C. Powder X-ray diffraction analysis of the sample after this treatment allows identification of the tetragonal phase $La_{0.5}Ba_{0.5}MnO_{2.5}$; this is apparent in a broadening of all peaks in the diffraction pattern.

Comments

(1) Accurate weighing of the chemicals and stirring of the reaction mixture prior to heating is crucial to obtain phase-pure samples. If this is not achieved, two crystalline impurities might be encountered: $La(OH)_3$ and $BaMnO_3$. Although the former can be removed by washing with dilute acid (5 wt % HNO_3 in water), the latter cannot be removed from the product mixture.

(2) TGA analysis under hydrogen shows that the material can be reduced to $La_{0.5}Ba_{0.5}MnO_{2.5}$, which has an ordered La/Ba arrangement. This indicates that the La^{3+} and Ba^{2+} ions are also ordered in layers in the hydrothermal $La_{0.5}Ba_{0.5}MnO_3$. This material is an unusual example of an A-site-ordered ternary perovskite; these phases can usually only be prepared in two steps via the reduced $Ln_{0.5}An_{0.5}MnO_{2.5}$ phase.[10] For ordered $La_{0.5}Ba_{0.5}MnO_3$, powder X-ray diffraction cannot be used to identify the A-site ordering since La^{3+} and Ba^{2+} are isoelectronic; this explains why the powder X-ray data can be indexed using a primitive cubic unit cell. Powder neutron diffraction must be used to verify the La/Ba ordering since La^{3+} and Ba^{2+} have differing neutron scattering lengths.

B. Preparation of $La_{0.5}Sr_{0.5}MnO_3$

Fresh solutions of 0.300 M $KMnO_4$, 0.350 M $MnSO_4$, and 0.350 M $La(NO_3)_3$ were prepared using volumetric flasks. 2.00 mL of the $MnSO_4$ solution were mixed with 1.43 mL of the $La(NO_3)_3$ solution and 0.0918 g of $SrSO_4$ in the Teflon™ liner of an autoclave with stirring. To this mixture was added 1.00 mL of the $KMnO_4$ solution, and finally 7.00 g of KOH pellets with continued stirring. The final composition of the mixture was $7Mn^{2+}$: $3MnO_4^-$: $5Sr^{2+}$: $5La^{3+}$: 1250 KOH : 3256 H_2O. The complete mixture was stirred for 30 min before sealing the Teflon™ liner, containing the reagents, in the stainless-steel autoclave. The autoclave was heated at 240°C for 24 h, followed by cooling at 2°C min^{-1} to room temperature. The solid product, a fine black powder, was isolated by suction filtration, washed with distilled water and acetone, and dried in air at 100°C for 24 h.

Characterization

Powder X-ray diffraction data (Fig. 2, Table 2) can be indexed using a primitive cubic unit cell with $a = 3.867(3)$ Å.

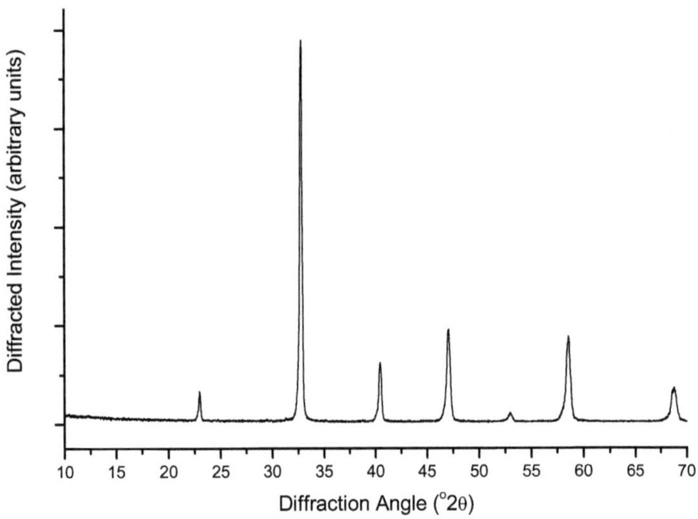

Fig. 2. Laboratory powder X-ray diffraction data from $La_{0.5}Sr_{0.5}MnO_3$ prepared by hydrothermal synthesis.

Iodometric titration, performed in the same way as for $La_{0.5}Ba_{0.5}MnO_3$, showed that the material has a composition close to the ideal stoichiometry $La_{0.5}Sr_{0.5}MnO_3$.

Thermogravimetric analysis recorded under an atmosphere of 10 % H_2 in Ar shows a mass loss of 3.9 % between 100 and 600°C. Powder X-ray diffraction analysis of the sample after this treatment showed that the material has collapsed into various mixed La/Sr/Mn oxide phases.

Table 2. Indexed powder X-ray diffraction data from $La_{0.5}Sr_{0.5}MnO_3$.

$2\theta_{obs}$*	I_{obs}	$2\theta_{calc}$	*hkl*
23.005	6.8	22.9977	001
32.7949	100.0	32.7498	011
40.4632	17.1	40.3980	111
47.0765	35.5	46.9935	002
52.9911	3.5	52.9435	012
58.5459	40.6	58.4580	112
68.7618	22.0	68.6439	202

* Cu $K_{\alpha1}/K_{\alpha2}$ radiation, recorded from 5-70 °2θ.

Comments

(1) As with $La_{0.5}Ba_{0.5}MnO_3$, accurate weighing and stirring of the reagents is essential to achieve a phase-pure sample. In this case, $La(OH)_3$ is the only crystalline impurity that might be encountered; any excess Sr^{2+} will remain in solution.

(2) The powder X-ray diffraction data for hydrothermal $La_{0.5}Sr_{0.5}MnO_3$ can be indexed on a cubic unit cell. This is in contrast to materials with the same composition prepared by ceramic synthesis. These processes result in a structure of lower symmetry (typically a mixture of orthorhombic and tetragonal polymorphs is produced [11]). If the hydrothermal material is subsequently fired at 1500°C, then transformation to the lower symmetry forms will take place. Unlike the barium-doped material, cubic $La_{0.5}Sr_{0.5}MnO_3$ does not exhibit A-site ordering: this is confirmed by powder X-ray diffraction and the fact that in the TGA experiment under H_2 the material collapses and undergoes phase separation into various Sr/Mn/La mixed oxides.

References

[1] C. N. R. Rao, J. Phys. Chem. B 2000, **104**, 5877.
[2] T. Seiyama, Catal. Rev. 1992, **34**, 281.
[3] N. Minh, J. Am. Ceram. Soc. 1993, **76**, 563.
[4] J. Spooren, A. Rumplecker, F. Millange, R. I. Walton, Chem. Mater. 2003, **15**, 1401.
[5] C. Bernard, C. Laberty, F. Ansart, B. Durand, Anal. Chim., Sci. Mater. 2003, **28**, 85.
[6] D. L. Zhu, H. Zhu, Y. H. Zhang, J. Cryst. Growth 2003, **249**, 172.
[7] J. Spooren, R. I. Walton, J. Solid State Chem. 2005, **178**, 1683.
[8] J. Spooren, R. I. Walton, F. Millange, J. Mater. Chem. 2005, **15**, 1542.

[9] C. Vázquez-Vázquez, M. C. Blanco, M. A. López-Quintela, R. D. Sánchez, J. Rivas, S. B. Oseroff, J. Mater. Chem. 1998, **8**, 991.

[10] F. Millange, V. Caignaert, B. Domengès, B. Raveau, E. Suard, Chem. Mater. 1998, **10**, 1974.

[11] P. M. Woodward, T. Vogt, D. E. Cox, A. Arulraj, C. N. R. Rao, P. Karen, A. K. Cheetam, Chem. Mater. 1998, **10**, 3652.

Lithium Manganese Oxide Prepared by Flux Methods

W. Tang, Q. Feng, X. Yang, R. Chitrakar and K. Ooi

Abstract Plate-like crystals of spinel-type lithium manganese oxide with sizes in the micrometer range were obtained in a flux system of LiCl-$Mn(NO_3)_2$. The crystals were grown by a dissolution-recrystallization mechanism. Polyhedral crystals with sizes in the millimeter range were grown in a flux system of LiCl-$Mn(NO_3)_2$ by an evaporation-recrystallization mechanism.

Classification

form:	single crystals
function:	selective adsorbent, catalyst, electrode material
preparation:	melting salt flux
composition:	$LiMn_2O_4$

Introduction

Spinel-type lithium manganese oxides are an attractive Li^+ adsorbent,[1] cathode materials for advanced lithium batteries,[2,3] and electrode materials for selective electroinsertion of Li^+.[4] Single crystals of lithium manganese oxide spinel may be used as cathodes for micro-sized rechargeable batteries or as lithium ion sensors. $LiMn_2O_4$ spinel has a structure with Li^+ at the 8a tetrahedral sites and Mn(III) and Mn(IV) at the 16d octahedral sites in a cubic closed-packed lattice of oxide ions.[5] The compound has some distinct characteristics, e.g. an easy conversion between Mn(III) and Mn(IV) and an easy Li^+ migration in the oxide lattice. These properties enable lithium manganese oxide spinels to have various composition with different oxidation states of Mn and with different Li/Mn ratios.[6] Lithium manganese oxide spinels are commonly prepared by solid-state reactions or sol-gel processing, which results in inhomogeneous polycrystalline materials with irregular morphology.[7]

Molten salt fluxes, an ionic non-aqueous environment, usually provide single crystals, or polycrystals with high crystallinity and distinct morphologies. Highly pure lithium manganese oxides without other cation contamination can be prepared by using Li-containing fluxes. These fluxes can be classified in four types according the mechanism of the reaction between the fluxes and manganese sources: (1) oxidizing, (2) non-oxide, (3) oxidic, but not oxidizing, and (4) no reaction. The nature of the fluxes and the resulting products are summarized in Table 1, where some results of mixed fluxes are also listed. The mixed fluxes have the merit of lowering the melting point.

Table 1. Manganese oxides obtained in different Li-containing molten salt fluxes.

Flux type	Li salt	m.p. [°C]	Product	Ref.
Oxidizing	$LiNO_3$	255	Spinel, β-MnO_2, Li_2MnO_3	[8,9]
	$LiClO_4$	236	β-MnO_2	unpublished
Non-oxide	LiCl	610	Spinel, Li_2MnO_3, $LiMnO_2$	[3,10, 11,12]
Oxidic, not oxidizing	LiOH	477	Li_2MnO_3	[12]
	Li_2CO_3	730		[12]
No reaction	Li_2SO_4	700	Mn_2O_3	[12]
	Li_2WO_4	742		unpublished
Mixed	$50LiBO_2 \cdot 50LiCl$	-	Spinel	[13]
	$31LiOH \cdot 69LiCl$	280	$LiMnO_2$ [a]	[14]
	$30LiF \cdot 70LiCl$	501	Li_2MnO_3	unpublished

[a] Under N_2 atmosphere

The lithium content, the mean oxidation state of manganese and oxygen content in the products are sensitive to the nature of the flux and heating conditions, including temperature, time, atmosphere, etc. LiCl and $LiNO_3$ fluxes give rise to various kinds of lithium manganese oxides compared with other fluxes. A LiCl flux is advantageous in the preparation of lithium manganese oxide crystals containing Mn(III), and a $LiNO_3$ flux in the preparation of those containing Mn(IV). Therefore, these two fluxes are most important for preparation of lithium manganese oxide spinels with different composition and morphologies.

In this contribution, a molten salt flux method for the synthesis of single crystals of spinel-type lithium manganese oxides using LiCl fluxes is described. Single crystals with millimeter and micrometer size are obtained in this non-oxide LiCl flux using $MnCl_2$ and $Mn(NO_3)_2$ as the manganese sources, respectively.

Materials

- Anhydrous LiCl, $Mn(NO_3)_2 \cdot 6H_2O$, and $MnCl_2 \cdot 6H_2O$, purchased from Wako Pure Chemical Industries, Ltd.
- An aqueous 38 wt% $Mn(NO_3)_2$ solution, prepared by dissolving $Mn(NO_3)_2$" $6H_2O$ in distilled water.
- An aqueous 2 M $MnCl_2$ solution, prepared by dissolving $MnCl_2$" $6H_2O$ in distilled water.
- High purity alumina crucible with 150 ml volume.

Safety an Disposal

Safety and handling instruction for the chemicals are found in the corresponding materials safety data sheets (MSDS). LiCl-$Mn(NO_3)_2$ and LiCl-$MnCl_2$ flux systems produce toxic gases of Cl_2 during heating. The gas should be exhausted into alkali solutions or adsorbents.

Procedures

A. Growth of $LiMn_2O_4$ Spinel Single Crystals in a LiCl-$Mn(NO_3)_2$ Flux [3]

Single crystals of $LiMn_2O_4$ with micrometer size are prepared in a LiCl-$Mn(NO_3)_2$ reaction system, in which LiCl is used as a flux and lithium source and $Mn(NO_3)_2$ as the manganese source. To obtain an uniform mixture, a 38% $Mn(NO_3)_2$ solution containing 10.6 g of $Mn(NO_3)_2$ and 50 g of LiCl are mixed in a beaker of 200 ml. The molar ratio of $Mn(NO_3)_2$ to LiCl in this mixture is 0.05. The mixture is dried at 120°C for 4 h, and then put into an alumina crucible (150 ml). The crucible is then heated in an electric furnace at 650, 750 or 850°C and the heating time is set in a range from 4 min to 24 h. The electric furnace is then allowed to cool to room temperature. The cooling progresses exponentially and takes about 10 or 50 min from 650 or 850°C to the temperature of 610°C (melting point of LiCl). The melt is then dispersed in 1 L of distilled water to dissolve the LiCl flux. After the flux is dissolved, single crystals of $LiMn_2O_4$ are collected by filtering. They are washed with distilled water and dried at 100°C. For the products obtained for 4 min heating, post-annealing after isolation of the crystals is carried out at 750°C for 4 h to promote the crystallization.

Characterization

The $LiMn_2O_4$ single crystals obtained at 650, 750, and 850°C for 4 min are identified as pure $LiMn_2O_4$ by powder X-ray diffraction analysis (Table 2). SEM images of the products obtained at 650, 750 and 850°C are shown in Fig. 1. Plate-like crystals are formed at 650, 750 and 850°C. The crystallite size increases with increasing temperature. The thickness of single crystals is less than 0.1 μm for the product at 650° and about 0.2 and 0.6 μm for those at 750 and 850°C, respectively. $LiMn_2O_4$ crystals are produced rapidly in a LiCl-$Mn(NO_3)_2$ flux system. The single spinel phase can be obtained at 650 and 750°C in a reaction time up to 24 h. After reaction at 650°C for 24 h, the crystal size of the plate-like crystal increases. Plate-like crystals are obtained at 750°C for 8 h, while the product obtained at 750°C for 24 h consists mainly of polyhedral crystals. The lithium and manganese contents in the products are determined by atomic absorption spectroscopy after dissolving the product in a mixture of H_2SO_4 and H_2O_2. The lattice constants, crystal size and chemical composition of the products obtained at different heating temperatures and times are summarized in Tables 2 and 3.

Table 2. Lattice constants (±0.01 Å), thickness and chemical composition of crystals obtained at 650, 750 and 850°C for 4 min in the LiCl-$Mn(NO_3)_2$ system.

Temperature [°C]	Lattice constant [Å]	Thickness [μm]	Composition
650	8.18	<0.1	$Li_{1.16}Mn_2O_{4.31}$
750	8.22	0.2	$Li_{1.08}Mn_2O_{4.10}$
850	8.22	0.6	$Li_{0.98}Mn_2O_{3.99}$
After post-annealing at 750°C			
650	8.21	<0.1	$Li_{1.18}Mn_2O_{4.28}$
750	8.24	0.2	$Li_{1.06}Mn_2O_{4.15}$
850	8.24	0.6	$Li_{1.00}Mn_2O_{4.02}$

Table 3. Lattice constants (±0.01 Å), thickness, and chemical composition of crystals obtained at 650 and 750°C for 2, 8 or 24 h in the LiCl-$Mn(NO_3)_2$ system.

Temperature [°C]	Time [h]	Lattice constant [Å]	Thickness [μm]	Composition
	2	8.23	0.2	$Li_{1.10}Mn_2O_{4.23}$
650	8	8.24	0.3	$Li_{1.08}Mn_2O_{4.17}$
	24	8.24	0.5	$Li_{1.04}Mn_2O_{4.12}$
	2	8.24	0.3	$Li_{1.06}Mn_2O_{4.10}$
750	8	8.24	0.7	$Li_{1.04}Mn_2O_{4.10}$
	24	8.24	2 [a]	$Li_{1.06}Mn_2O_{4.12}$

[a] Average size

Fig. 1. SEM images of $LiMn_2O_4$ crystals obtained at 650, 750 and 850°C for 4 min in LiCl-$Mn(NO_3)_2$ system.

Comments

(1) The product obtained at 950°C for 4 min consists mainly of $LiMn_2O_4$ crystals with a minor amount of Mn_3O_4.
(2) The post-annealing changes the oxygen contents in the crystals, without changing the crystal morphology.
(3) The single crystals are grown by a dissolution-recrystallization mechanism in this flux system, because the crystal size increases distinctly with increasing the heating temperature and time. LiCl is a neutral salt, it is expected to have high solubility for manganese and to stabilize Mn^{3+} in the LiCl melt, in contrast to oxygen-containing salts.[8,15] $Mn(NO_3)_2$ as the manganese source shows higher reactivity than that of solid manganese oxides in this flux system.

B. Growth of $LiMn_2O_4$ Spinel Single Crystals in a LiCl-$MnCl_2$ Flux [11]

A 2 M aqueous $MnCl_2$ solution is mixed with LiCl (50 g) to prepare a mixture of LiCl and $MnCl_2$. The Mn content in the mixture is adjusted in a range of 8 to 12 mmol using 4 to 6 ml of the $MnCl_2$ solution. The mixture is then dried at 180°C for 3 h. After grinding, the mixture is placed in a pure alumina crucible (150 ml in volume), and then covered with a 40 g layer of LiCl on the top of the mixture to prevent sudden evaporation of LiCl and oxidation of $MnCl_2$ by air during the heating process. The crucible is covered with a pure alumina lid, and then heated at 750°C for 58 h in an electric muffle oven. The electric furnace is then allowed to cool to room temperature. Polyhedral $LiMn_2O_4$ single crystals, black-colored with a glossy surface, are formed on the wall of the crucible above the LiCl melt (i.e. at the crucible wall / LiCl melt / air interface), arranged as a 5 mm wide band. The melt is dissolved in 1 ml of distilled water, and the $LiMn_2O_4$ crystals are collected by filtration. They are washed with distilled water and then dried at 100°C.

Characterization

The products are characterized as described above. A X-ray powder analysis shows that the products are single spinel phase of $LiMn_2O_4$. The chemical formula of the spinel phase obtained at 8 mmol of Mn is $Li_{1.03}Mn_{1.97}O_4$ by chemical analysis, which is very close to the theoretical formula of $LiMn_2O_4$. The yield of single crystals is 0.18 g and 0.26 g at 8 and 12 mmol of Mn content, respectively, and increased with increasing Mn content. Octahedral and rectangular $LiMn_2O_4$ single crystals are obtained at 8 and 12 mmol of Mn content. These crystals have smooth surfaces with sizes larger than 0.1×0.1×0.1 mm. Fig. 2 shows a SEM image of the crystals obtained at 8 mmol of Mn content.

Fig. 2. SEM image of $LiMn_2O_4$ single crystals obtained at 8 mmol of Mn content in LiCl-$MnCl_2$ system.

Comments

(1) The product consists of mainly $LiMn_2O_4$ with minor amount of Li_2MnO_3 or Mn_2O_3 when the Mn content is less than 6 mmol, or larger than 18 mmol, respectively. The crystal shape also changes with the Mn content. Crystals larger than 0.9×0.5×0.5 mm can be obtained at 18 mmol of Mn content, but the crystals are intergrown.
(2) The single crystals grow by an evaporation-recrystallization mechanism in this flux system, because the crystals are formed on the wall of the crucible above the LiCl-$MnCl_2$ melt. The LiCl-$MnCl_2$ mixture evaporates heavily in the flux process at the temperature over its melting point.

References

[1] V. V. Vol'khin, G. V. Leont'eva, S. A. Onolin, Neorg. Mater. 1973, **6**, 1041. A.Umeno, Y.Miyai, N. Takagi, R. Chitrakar, K. Sakane, K. Ooi, Ind. Eng. Chem. Res. 2002, **41**, 4281.
[2] M. Thackeray, W. I. F. David, P. G. Bruce, J. B. Goodenough, Mater. Res. Bull. 1983, **18**, 461.
[3] W. Tang, X. Yang, Z. Liu, K. Ooi, J. Mater. Chem. 2002, **12**, 2991.

[4] H. Kanoh, K. Ooi, Y. Miyai, S. Katoh, Langmuir 1991, **7**, 1841.
[5] J. C. Humter, J. Solid State Chem. 1981, **39**, 142.
[6] G. Blasse, Philips Res. Rep., Suppl. 1964, No.3, 1.
[7] X. M. Shen, A. Clearfied, J. Solid State Chem., 1986, **64**, 270. B. Ammundsen, D. J. Jones, J. Roziere, G. R. B Burns, Chem. Mater., 1997, **9**, 3236.
[8] X. Yang, H. Kanoh, W. Tang, K. Ooi, J. Mater. Chem. 2000, **10**, 1903.
[9] X. Yang, W. Tang, Z. Liu, Y. Makita, K. Ooi, J. Mater. Chem. 2002, **12**, 489. W. Tang, Y. Ogo, H. Kanoh, X. Yang, K. Ooi, Proc. Joint 6th Int. Symp. on Hydrothermal Reaction & 4th Int. Conf. on Solvo-Thermal Reactions, Eds. K. Yanagisawa, Q. Feng, 2000, p.419.
[10] W. Tang, H. Kanoh, K. Ooi, Chem Lett. 2000, 216. W. Tang, H. Kanoh, X. Yang, K. Ooi, Chem. Mater. 2000, **12**, 3271.
[11] W. Tang, X. Yang, H. Kanoh, K. Ooi, Chem. Lett. 2001, 524.
[12] P. Strobel, J. P. Levy, J. C. Joubert, J. Cryst. Growth 1984, **66**, 257.
[13] J. Akimoto, Y. Takahashi, Y. Gotoh,, S. Mizuta, Chem. Mater. 2000, **12**, 3246.
[14] W. Tang, H. Kanoh, K. Ooi, J. Solid State Chem. 1999, **142**, 19.
[15] X. Yang, W. Tang, H. Kanoh, K. Ooi, J. Mater. Chem. 1999, **9**, 2683.

Nanoscale Magnesium Oxide

F. Khairallah and A. Glisenti

Abstract Magnesium oxide (MgO) with a crystallite size of 9±1-2 nm and a surface area of 65 m^2g^{-1} was synthesized by precipitation of magnesium oxalate in the presence of tetraethylammonium hydroxide as a peptization agent, followed by calcination under controlled conditions.

Classification

form:	nanocrystalline powder
function:	catalyst, catalyst support
preparation:	precipitation from aqueous solution, calcination
composition:	MgO

Introduction

Magnesium oxide is one of the most important industrial magnesium compounds used for a variety of applications (refractory materials, pharmaceuticals, waste remediation, glass industry and catalysis).[1-2]

While the refractory market typically uses conventional MgO powders, there is an emerging market for MgO with special characteristics such as higher surface area, finer crystallites, higher purity and enhanced chemical activity for catalytic applications. It has been reported that nanoscale MgO exhibits a different reactivity than conventionally prepared MgO.[3] Previous work [4-6] has shown that nanocrystalline MgO exhibits remarkable capacities and rates of adsorption, primarily because of the unique reactivity. This is attributed to the large surface areas and the enhanced surface reactivity which is due to the unusual crystal shapes with a large proportion of coordinatively unsaturated edge/corner surface sites as well as defect sites being more reactive toward incoming adsorbates.[4-5]

MgO is synthesized by several processes, the most common being the calcination of $Mg(OH)_2$, such as brucite, precipitation from sea water or decomposition of $MgCO_3$, such as magnesite or dolomite. MgO prepared by different routes has different surface morphologies, and thus the catalytic activity and selectivity may vary.[3,6-7] Other factors, such as temperature, treatment time, pH, gelling agent and heating atmosphere affect greatly the activity of the final product.[3,8]

Various preparative routes were found to produce nanoscale MgO including sol-gel methods,[2, 6, 9-14] the hydrothermal method,[15] flame spray pyrolysis, [16] laser vaporization,[5] chemical gas phase deposition,[17] combustion aerosol synthesis,[18] aqueous wet chemical methods,[19] surfactant methods,[20] etc. Some of these procedures require specific instrumentation and relatively meticulous experimental procedures. In this contribution a simple and reproducible method for the synthesis of nanoscale MgO by a wet chemical method is described which involves the precipitation of magnesium oxalate.

Materials

- $Mg(NO_3)_2.6H_2O$, Strem 99%
- Oxalic acid dihydrate, Aldrich 99%
- Aqueous ammonium hydroxide solution, Aldrich, 30%
- Aqueous tetraethyl ammonium hydroxide, Aldrich 20 wt%
- Deionized water

Safety and Disposal

Safety and handling instructions for the chemicals are found in the corresponding materials safety data sheets (MSDS). Standard safety precautions are required for operating with NH_4OH solution and oxalic acid.

Procedures

All the reagents were used as received without further purification.

A clear solution was prepared by dissolving 7.435 g (29 mmol) of $Mg(NO_3)_2.6H_2O$ in 50 ml of deionised water in a 150 mL beaker. The pH was raised to pH 8-9 by the dropwise addition of ammonium hydroxide solution (2-3 drops each). The stirred solution was gelled by the addition of an equimolar amount of oxalic acid (3.7 g, 29 mmol, dissolved in minimum amount of deionised water). To facilitate gelation/precipitation, tetraethylammonium hydroxide was added as a peptization agent in 0.5 ml portions (total approx. 2 mL) along with 0.5 ml portions of ammonium hydroxide (total approx. 1 mL), approx. 3-4 min were allowed for before adding another portion, this was continued until the solution started to become turbid (indicating the onset of precipitation). The precipitation step should take 30-45 min with continuous calm stirring. A very fine precipitate was obtained.

The precipitate was filtered using a Buchner funnel (water vacuum suction) and washed three times with 50 ml portions of deionised water each. The filtered sample was then dried in an oven at 120°C for 3 h (from room temperature to 120°C in 1 h), crushed using a mortar and subsequently calcined in a porcelain crucible using a muffle furnace in air for approx. 4 h at 550°C (from room temperature to 550°C in 2 h). The obtained product was a finely divided colorless powder.

Characterization

DRIFT spectra were collected using a Bruker IFS 66 spectrometer (accumulating 128 scans at a resolution of 4 cm^{-1}) and were displayed in Kubelka-Munk units [21-22]. To carry out the DRIFT spectroscopy measurements, ca. 50 mg of the sample were transferred in the sample cup of a low temperature reaction chamber (CHC) installed in the Praying Mantis™ accessory for diffuse reflection spectroscopy (Harrick Scientific Corporation) and fitted with ZnSe windows. The powder was kept in nitrogen flow to eliminate water traces until a stable IR spectrum was obtained (ca. 2 h). The spectra were collected as a function of the temperature. The sample annealed at 550°C had the following IR signals: 3760-2800 cm^{-1} (broad, ν_{OH} adsorbed moisture and ν_{OH} $Mg(OH)_2$),[23-25] 1750-1300 cm^{-1} (broad, $\nu_{C=O}$ carbonate species [mono- and bidentate, bicarbonate] as well as to bending H_2O).[25-27]

Note: The moisture and carbonate IR signals disappear at temperatures above 350°C.

XP spectra were recorded using a Perkin-Elmer PHI 5600*ci* spectrometer with a monochromated Al-Kα source (1486.6 eV) working at 300 W. The working pressure was less than 1×10^{-6} Pa. The spectrometer was calibrated by assuming the binding energy (BE) of the Au $4f_{7/2}$ line at 84.0 eV with respect to the Fermi level. Extended spectra (survey) were collected in the range 0-1400 eV (187.85 eV pass energy, 0.4 eV step, 0.05 s $step^{-1}$). Detailed spectra were recorded for the following regions: C1s, O1s, Mg1s and Mg2p (11.75 eV pass energy, 0.1 eV step, 0.1 s $step^{-1}$). The standard deviation in the BE values of the XPS line is 0.10 eV. The atomic percentage (element concentration), after a Shirley type background subtraction [28], was evaluated using the PHI sensitivity factors [29]. To take into consideration charging problems, the C1s peak at 285.0 eV was considered [30]. The sample for the XPS analysis was processed as a pellet by pressing the sample powder at ca. 7×10^{6} Pa for 10 min; the pellet was then evacuated for 12 h at ca. 1×10^{-3} Pa. Alternatively the powder sample was put directly on a carbon double sided adhesive tab for UHV. XP peak positions: 1303.5 eV (Mg1s), 49.7 eV (Mg2p), 530.4 and 532.1 eV (O1s). The peak positions are in agreement with the values expected for MgO [30-32]. XP atomic percentages: 45.6% (Mg1s), 54.4 % (O1s); O/Mg ratio: 1.19.

Note : Hydroxylation and carbonatation are the reason for not obtaining an O/Mg ratio of 1.

XRD patterns were obtained using a Bruker D8 Advance diffractometer with Bragg-Brentano geometry using Cu K_α radiation (40 kV, 40 mA, λ= 0.15406 nm). Scans were performed over the 2 θ range from 25 to 85°. The diffraction angles 2 θ (*relative intensities*) and the correspondent reflection planes: 36.98 (*10*)-111, 42.96 (*100*)-200, 62.32 (*44*)-220, 74.70 (*5*)-311, 78.56 (*11*)-222 match magnesium oxide reference pattern number 78-0430. An average crystallite size of 9 nm ± 1-2 nm was calculated applying Scherrer's formula with K = 0.9 (a constant dependent on the crystallite shape [19,33]). All the diffraction peaks were taken into consideration and an average was taken.

Thermogravimetric Analysis (TGA) was carried out in a controlled atmosphere using the Simultaneous Differential Technique (SDT) 2960 of TA Instruments. The thermograms were recorded with a 5°C min^{-1} heating rate in dry air and in nitrogen flow from room temperature to 800°C.

In dry air flow: 37-180°C (6.5% loss): desorption of adsorbed moisture. 205-410°C (12.3% loss): conversion of some residual $Mg(OH)_2$ into MgO and elimination of carbonate species.[6,34-35] 500-600°C (1.1% loss): elimination of some strongly bound water molecules deeply entrapped in the MgO lattice and elimination of carbonate species. In nitrogen flow: 35–187°C (6.5% loss). 206–412°C (12.1% loss). 498–605°C (1.1% loss).

Specific surface area 65.1±1.5 m^2g^{-1} (determined by N_2 sorption using BET method). The sample was outgassed for 4 h under vacuum at 100°C prior to analysis.

Comments

(1) In this procedure the addition of the peptization agent (tetraethylammonium hydroxide) is an important step, which should be done carefully. This is also the case for the ammonium hydroxide addition for raising the pH. The precipitation step depends on the addition of these two compounds where a sudden precipitation could occur resulting in a non-homogeneous and agglomerated precipitate.

(2) The temperature history followed is important for obtaining a homogenous product. Temperatures below 500°C do not assure the complete conversion of the precipitate into the desired MgO. Temperatures above 600°C could result in an increase of the particle size and a reduction of the surface area.

(3) All analyses should be performed after drying the sample overnight at 120°C. Otherwise the obtained results would be different due to a higher proportion of adsorbed moisture.

References

[1] R. E. Kirk, D. F. Othmer, M. Grayson, D. Eckroth, Kirk-Othmer Encyclopaedia of Chemical Technology; John Wiley and Sons, 1981, **14**.
[2] B. Q. Xu, J. M. Wei, H. Y Wang, K. Q. Sun, Q. M. Zhu, Catal. Today 2001, **68**, 217.
[3] H. C. Liang Septimus; I. D. Gay, Langmuir 1985, **1**, 593.
[4] A. Khaleel, P. N. Kapoor, K. J. Klabunde, Nanostruct. Mater. 1999, **11**, 459.
[5] M. El-Shall, W. Slack, W. Vann, D. Kane, D. Hanley, J. Phys. Chem. 1994, **98**, 3067.
[6] F. Khairallah, A. Glisenti, J. Mol. Catal. A 2007, **274**, 137.
[7] K. J. Klabunde, J. Stark, O. Koper, C. Mohs, D. G. Park, S. Decker, Y. Jiang, I. Lagadic, D. Zhang, J. Phys. Chem. 1996, **100**, 12142.
[8] V. Štengl, S. Bakardjieva, M. Maříkovà, P. Bezdička, J. Šubrt, Mater. Lett. 2003, **57**, 3998.
[9] J. Jiu, K. Kurumada, M. Tanigaki, M. Adachi, S. Yoshikawa, Mater. Lett. 2003, **58**, 44.
[10] A. V. Chadwick, I. J. F. Poplett, D. T. S.Maitland, M. E. Smith, Chem. Mater. 1998, **10**, 864.
[11] X. Bokhimi, A. Morales, M. Portilla, A. García-Ruiz, Nanostruct. Mater. 1999, 12, 589.
[12] X. Bokhimi, A. Morales, J. Solid State Chem. 1995, **115**, 411.
[13] T. López, R. Gómez, J. Navarrete, E. López-Salinas, J. Sol-Gel Sci. Technol. 1998, **13**, 1043.
[14] S. Utamapanya, K. J. Klabunde, J. R. Schlup, Chem. Mater. 1991, **3**, 175.
[15] Y. Ding, G. Zhang, H. Wu, B. Hai, L. Wang, Y. Qian, Chem. Mater. 2001, **13**, 435.
[16] R. M. Laine, C. R. Bickmore, D. R. Treadwell, K. F. Waldner, US 5,958,361, 1999.
[17] J. S. Matthews, O. Just, B. Obi-Johnson, W. S. Rees Jr., Chem. Vap. Deposition 2000, **6**, 129.
[18] J. J. Helble, J. Aerosol Sci. 1998, **29**, 721.
[19] A. Bhargava, J. Alarco, I. Mackinnon, D. Page, A. Ilyushechkin, Mater. Lett. 1998, **34**, 133.
[20] P. Talbot, J. A. Alarco, G. A. Edwards, WO 02/42201 (Cl. C01B13/36), 2002.
[21] P. Kubelka, F. Z. Munk, Tech. Phys. 1931, **12,** 593.
[22] G. Kortum, Reflectance Spectroscopy, Springer, New York, 1969.
[23] R. I. Razouki, R. SH. Mikhail, J. Phys. Chem. 1958, **62**, 920.
[24] H. W. Van der Marel, H. Beutelspacher, Atlas of Infrared Spectroscopy of Clay Minerals and Their Admixtures, Elsevier Scientific Publishing Co., New York, 1976.
[25] M. A. Aramendía; V. Boráu, C. Jiménez, A. Marinas, J. M. Marinas, J. A. Navío, J. R. Ruiz, F. J. Urbano, Colloids Surf. A 2004, **234**, 17.
[26] J. V. Stark, D. G. Park, I. Lagadic, K. J. Klabunde, Chem. Mater. 1996, **8**, 1904.
[27] www.lsbu.ac.uk/water/vibrat.html
[28] D. A. Shirley, Phys. Rev. B 1972, **5**, 4709.
[29] J. F. Moulder, W. F. Stickle, P. E. Sobol, K. D. Bomben, Handbook of X-Ray Photoelectron Spectroscopy, J. Chastain Ed., Physical Electronics: Eden Prairie, MN, 1992.
[30] D. Briggs, J. C. Riviere, Practical Surface Analysis, D. Briggs, M. P. Seah, Eds., Wiley Interscience, New York, 1983.
[31] S. Ardizzone, C. L. Bianchi, B. Vercelli, Colloids Surf. A 1998, **144**, 9.
[32] D. K. Aswal, K. P. Muthe, S. Tawde, S. Chodhury, N. Bagkar, A. Singh, S. K. Gupta, J. V. Yakhmi, J. Cryst. Growth 2002, **236**, 661.
[33] H. P. Klug, L. E. Alexander, X-ray Diffraction Procedures**,** 2nd ed., Wiley Interscience, New York, 1974.
[34] K. Itatani, A. Itoh, F. S. Howell, A. Kishioka, M. Kinoshita, J. Mater. Sci. 1993, **28**, 719.
[35] M. Crisan, A. Jitianu, D. Crisan, M. Balasoiu, N. Dragan, M. Zaharescu, J. Optoelectr. Adv. Mater. 2000, **2**, 339.

References

[1] R. E. Kirk, D. F. Othmer, M. Grayson, D. Eckroth, Kirk-Othmer Encyclopedia of Chemical Technology, John Wiley and Sons, 1981, 14.
[2] R. O. Xu, … W. Wei, H. … Wang, … Sun, Q. M. Zhu, Catal. Today … 117
[illegible]

Nanostructured Pt-doped Tin Oxide Films

R. Scotti, C. Canevali, M. Mattoni, F. Morazzoni, L. Armelao and D. Barreca

Abstract Nanostructured Pt-doped SnO_2 thin films are obtained by a sol-gel route using tetra(*tert*-butoxy)tin(IV) and bis(acetylacetonato)platinum(II) as precursors. Acetylacetone was added as a chelating agent to control the hydrolysis and condensation of tin alkoxide and, consequently, the gelation rate. This procedure allows obtaining a sol phase viscosity suitable for the spin-coating process.

Classification

form:	nanocrystalline thin films
function:	gas sensing
preparation:	sol-gel processing
composition :	Pt-doped SnO_2

Introduction

SnO_2 is the most extensively used material in solid state sensing devices for reducing gases (*e.g.* H_2, CO, NO, C_2H_5OH and CH_4) [1,2] based on the measurement of electrical sensitivity $S= R_{air}/R_{gas}$, where R_{air} and R_{gas} are the resistance in air and in the presence of the analyte gas, respectively. The sensing response is determined by the reactivity of point defects and chemisorbed oxygen species, as well as by the oxide microstructure and surface morphology.[3] According to the currently accepted gas sensing mechanism, nanostructured films improve the sensor performances due to high surface-to-volume ratio. As a matter of fact, in this case each SnO_2 grain is included as a whole in the space charge region and reactions with the surrounding gases affect the electron transport throughout the particle.[3]

The sensing response as well as the selectivity towards different reducing gases can be further improved by adding small amounts of suitable noble metals. The origin of the promoting effect due to the metal is still under discussion. Two mechanisms were proposed:[4,5] (i) the catalytic activation and spillover of the gas, H_2, that reacts more easily with oxygen adsorbates; (ii) the electronic sensitization due to the electron transfer from the oxide to O_2 throughout the doping metal which, in turn, switches between different oxidation states. As a consequence, the amount and dispersion of the metal could affect both the amplitude and the reproducibility of the electrical response. Studies on polycrystalline samples of transition metal-doped SnO_2 [6,7] aimed at investigating the role of metal location in the SnO_2 lattice with respect to the improvement of the electronic exchange between the surrounding atmosphere and the metal/oxide system.

Different techniques have been employed to optimize the preparation and the sensitivity of metal-doped SnO_2 films. Among them, the sol-gel route [2,8,9] represents a low-temperature approach, which allows a fine control of the layer composition, microstructure and morphology.[10] Nanostructured metal-doped SnO_2 sol-gel coatings were already prepared by our group and attention was focused on the relationships between the film properties and the electrical sensitivity.[11-14]

Materials

- Bis(acetylacetonate) platinum (II), $Pt(acac)_2$, Aldrich, 97%.
- Ethanol, HPLC grade reagent; water Mill-Q
- Acetylacetone, Hacac, Aldrich, 99+%.

Safety and Disposal

Safety and handling instructions for the chemicals are found in the corresponding materials safety data sheets (MSDS). Acetylacetone is an inflammable liquid and a mild irritant to the skin and the mucous membrane. $Sn(OBu^t)_4$ *is moisture sensitive.* $Pt(acac)_2$ *is an irritant and possible teratogen agent, and neurologically hazardous.*

Procedures

The sol-gel route for nanostructured (average grain size $\leq$6 nm) Pt-doped SnO_2 thin films with a 0.025 Pt:Sn molar ratio, using tetra(*tert*-butoxy)tin(IV) and bis(acetylacetonato)platinum(II), $Pt(acac)_2$, as precursors is described. Acetylacetone is added as a chelating agent to control the hydrolysis and condensation of tin precursor [10] and, consequently, the gelation rate. Sol phases with suitable viscosity for spin-coating deposition are thus obtained. In principle, the described procedure can be used to prepare SnO_2 films doped with any transition metal from acetylacetonate derivatives soluble in ethanol-acetylacetone (EtOH-Hacac) solutions.

A. Preparation of Tetra(tert-butoxy)tin [15]

A 2.0 M solution of $SnCl_4$ in *n*-pentane (250 mL) was prepared in a four-necked, round-bottomed 1 l flask fitted with a N_2 inlet, reflux condenser, pressure-equalized addition funnel and mechanical stirrer. The solution was cooled to 273 K, and Et_2NH_2 (90 mL) dissolved in *n*-pentane (40 mL) was added dropwise over a period of 2 h. An exothermic reaction occurred with the formation of a white precipitate. Further Et_2NH_2 (120 mL) dissolved in *n*-pentane (100 mL) was added dropwise. The mixture was stirred at room temperature for 4 h, then cooled to 273 K, and four 50 mL portions of *t*-BuOH were added over 10 min for each portion. When the addition was complete, the mixture was stirred for 14 h at room temperature. The stirrer was stopped and the mixture allowed to stand for 2 h and was then filtered. The remaining white solid was transferred to a Soxhlet extractor and extracted with *n*-pentane (300 mL). The pale-yellow *n*-pentane solutions were combined and the volatile components were removed *in vacuo* (10^{-2} Torr) to give a pale yellow solid, which was then sublimed at 10^{-2} Torr and 313 K onto a cold finger maintained at 195 K. Colorless crystals of $Sn(OBu^t)_4$ were thus obtained.

B. Preparation of the Sol for Coating

The coating sol was prepared under nitrogen atmosphere by mixing

- 2.50 mL of a solution of $Sn(OBu^t)_4$ (360 mg·ml^{-1}) in anhydrous EtOH-Hacac 3.85:1 v:v (corresponding to 2.19 mmol of Sn),
- 3.00 mL of a $Pt(acac)_2$ solution (7.17 mg·mL^{-1}) in EtOH-Hacac 1:1 v:v (corresponding to $5.47{\cdot}10^{-2}$ mmol of Pt),
- 1.52 mL of EtOH and 0.48 mL of Hacac,
- 1.00 mL of an ethanol-water solution 4:1 v:v (corresponding to 11.1 mmol of H_2O).

The solution was transferred into a cylindrical container (section area $\approx$ 5 cm^2) connected with a modified Ubbelohde capillary viscometer (Fig. 1) and kept into a thermostatic chamber at 308±1k. After 24 h, additional 0.10 mL of the ethanol-water solution (4:1 v:v) were introduced (corresponding to 1.11 mmol of H_2O) and the addition was repeated every 24 h until the sol phase viscosity was 2.5 cSt (1 cSt = 10^{-6} m^2·s^{-1}), a suitable value for spin coating deposition. The cinematic viscosity was determined using the Ubbelhode viscometer by measuring the flow time of a precise solution volume through a capillary tube under gravity force. The viscometer was modified to avoid solvent evaporation, which could modify the condensation kinetics and affect the viscosity control during sol phase aging. Two vessels sealed on the top of the viscometer (A in Fig. 1) are connected by a three-way valve (B), which permits pressure equalization in the system. The solution was aspirated by a syringe through the septum (C) up to the top of the capillary (D) for the measurement while keeping the system closed.

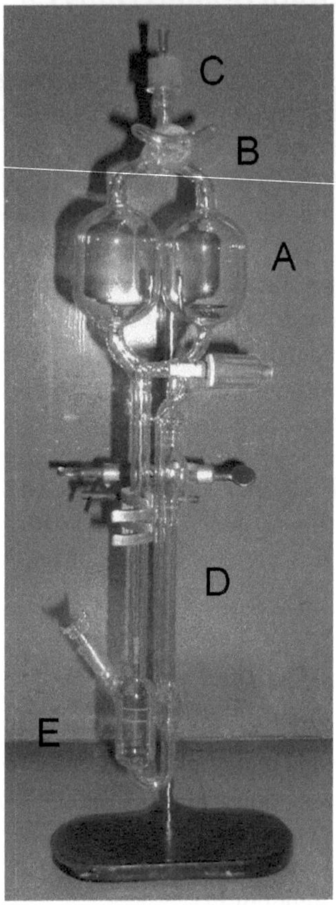

Fig. 1. Ubbelhode viscometer: (A) vessels; (B) three-ways valve; (C) septum; (D) capillary; (E) container of the solution.

The viscosity increase was controlled by stepwise addition of small amounts of the ethanol-water mixture. The viscosity was measured immediately after the preparation of the sol phase (1.1 cSt) and 24 h later (1.4 cSt). After that, the viscosity was measured every 24 h, i.e. after each additional addition of the ethanol-water mixture (Fig. 2). Small differences in viscosity (±0.3 cSt) were observed for different sol-phase preparations after the same number of ethanol-water additions. Viscosity increased slowly up to ≈2.5 cSt after five additions of 0.10 mL of ethanol-water mixture each, followed by a faster increase (4.7 cSt after the seventh addition of 0.10 mL of ethanol-water mixture). However, complete gelation did not occur even 360 h after the sol preparation, in spite of the high viscosity (>10 cSt). The stepwise addition of ethanol-water instead of addition of the whole amount allowed lowering and controlling the hydrolysis and condensation reaction rate. In fact, the addition of 0.50 mL of ethanol-water mixture (equivalent to five successive additions of 0.10 mL every 24 h) to the same amount of sol caused complete gelation within 24 h.

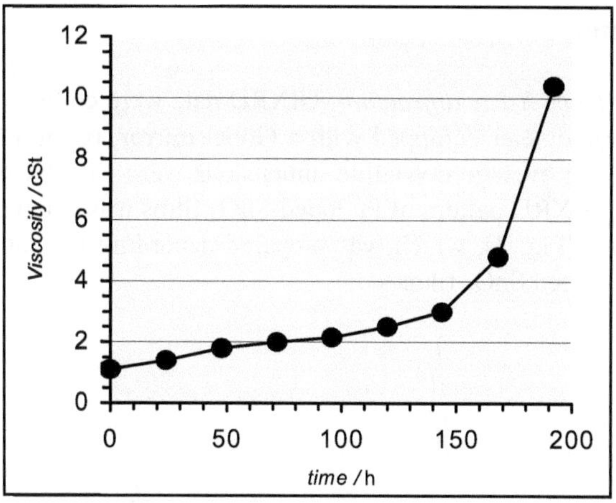

Fig. 2. Plot of viscosity *vs.* time during the addition of the ethanol-water mixture to the EtOH solution of $Sn(OBu^t)_4$ and $Pt(acac)_2$ (Pt:Sn 0.025 molar ratio). An amount of 0.1 mL of ethanol-water was added after each measurement (see text).

C. Film Deposition and Annealing

Films were deposited by spin-coating in air on silica glass slides. The substrate dimensions ranged between 10×10 mm and 40×40 mm. Before deposition, the glass slides were cleaned according to the following procedure:

- washing in trichloroethylene at boiling temperature (5 min),
- washing in acetone at boiling temperature (5 min),
- fast washing in water at room temperature,
- storage in ethanol.

The spinning profile consisted of three steps:

- 100 rounds per minutes (rpm) for 2 s (the spin rate was gained in 1 s),
- 200 rpm for 2 s (the spin rate was gained in 1 s),
- 2000 rpm for 60 s (the spin rate was gained in 3 s).

After deposition, the films were dried at room temperature and finally annealed at 673 K or 973 K in a stream of air (80 $cm^3 \cdot min^{-1}$) for 2 h. Annealing was performed in an oven equipped with a quartz chamber connected to a gas line. Thermal treatment in air allowed a clean decomposition of the acetylacetonate precursor (see XPS analysis).[7]

Multiple depositions (two or three) on the same substrate were also performed to obtain thicker coatings. In this case, each deposited layer was annealed at 673 K in a stream of air (80 $cm^3 \cdot min^{-1}$) for 2 h before additional depositions. The film thickness, measured by a Tencor P-10 surface profiler, was 80, 160 and 240 nm for single, double and triple deposition, respectively, after annealing at 673 K.

Characterization

Glancing incidence X-ray diffraction: GIXRD data were collected on a Bruker D8 Advance diffractometer equipped with a Göbel mirror, using a CuK_α source (40 kV, 40 mA). The average crystallite dimensions were calculated by the Scherrer equation. The GIXRD pattern of Pt doped-SnO_2 films was compared with that of a pure SnO_2 film (Fig. 3), which was prepared according the same procedure described for Pt-doped SnO_2 films.

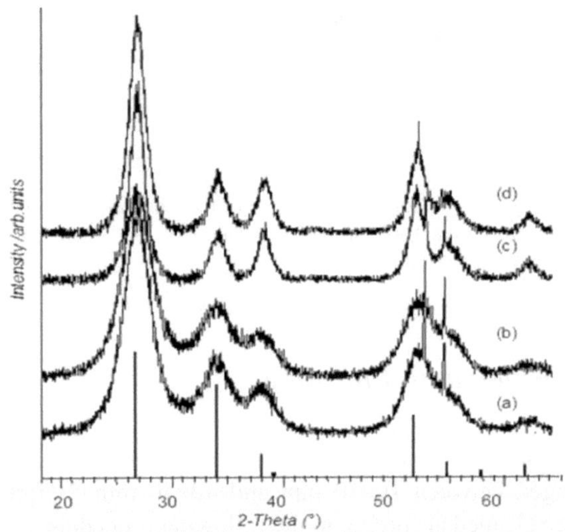

Fig. 3. GIXRD spectra (incidence angle 1°) of Pt-doped SnO_2 and SnO_2 films annealed at 973 K (a and b) and annealed at 673 K (c and d). The diffraction pattern of cassiterite is also reproduced for comparison.

The main structural information that can be inferred by these measurements is that:

(1) The SnO_2 cassiterite phase was observed both in pure and doped films after annealing at 673 or 973 K.[11]
(2) The average particle size of the cassiterite crystallites was dependent on the annealing temperature. Average dimensions of ≈3 and 6 nm were obtained for both pure and doped SnO_2 films after annealing at 673 and 973 K, respectively.[11]
(3) GIXRD measurements performed at different incident angles (0.2°–5°) excluded possible preferred orientations of the crystalline grains in the film.[11]
(4) No crystalline phases related to platinum were detected except for measurements performed at 0.2° incident angles. In this case, a diffraction peak due to a surface segregated PtO_x phase was detected. This effect possibly explains why the intensity ratio for the cassiterite peaks at $2\theta = 52°$ and 54.5° was dif-

ferent in Pt-doped SnO_2 sample annealed at 973 K with respect to the ideal SnO_2 pattern.[11]

X-ray photoelectron spectroscopy: XPS spectra were recorded on a Perkin-Elmer Φ 5600-ci spectrometer using non-monochromatized Al-K_α radiation (1486.6 eV) and a working pressure of $<5 \cdot 10^{-8}$ Pa. The standard deviation for the binding energy (BE) values was ±0.15 eV. BEs were corrected for the charging effects, assigning to the C1s line of adventitious carbon a value of 285.0 eV. The chemical composition and element distribution along the film thickness was evaluated by in-depth analysis. Depth profiles were carried out by Ar^+ sputtering at 2.5 keV with an Argon partial pressure of $5 \cdot 10^{-6}$ Pa. Under these conditions the sputtering rate was around 10 Å·min^{-1}. Samples were introduced directly into the XPS analytical chamber by a fast entry lock system. The main chemical information obtained by XPS analysis are:[11]

(1) *Oxidation state of the doping metal.* Pt-SnO_2 films showed the presence of Pt (IV) centers (BE Pt4$f_{7/2}$ ≅ 74.6 eV) after annealing in air at 673 K (Fig. 4). Changes in Pt oxidation state, due to exposure to different gases, was also investigated. As an example, the reactions of Pt-SnO_2 films with a CO (600 ppm)/Ar atmosphere [11] resulted in the reduction of Pt(IV) to Pt(II) (BE Pt4$f_{7/2}$ ≅ 72.7 eV) at 373 K and Pt(0) (BE Pt4$f_{7/2}$ ≅ 71.7 eV) at 673 K (Fig. 4).

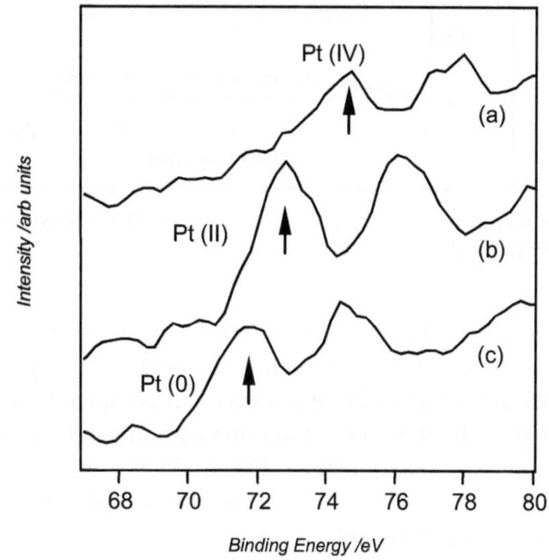

Fig. 4. XPS Pt4f spectra of the Pt-doped SnO_2 film: (a) annealed in air at 673 K; (b) annealed in air at 673 K and then treated with CO(600 ppm)/Ar flow at 373 K; (c) annealed in air at 673 K and then treated with CO(600 ppm)/Ar stream at 673 K.

(2) *Composition of the film and in - depth distribution of the doping metal.* The elemental composition was calculated from the intensities of Pt4f, Sn3d (BE =

486.7 eV, characteristic of tin oxide [16]) and O1s (BE = 530.6 eV, typical for cassiterite [16]) lines.

XPS in depth-analysis (Fig. 5) showed that Pt:Sn atomic ratio increased from the outer to the inner layers of the film. The surface Pt:Sn value was lower than expected from the sol phase composition (Pt:Sn 0.025) and progressively increased towards the film-substrate interface region. The O:Sn ratio was in agreement with the stoichiometric composition at the film surface, whereas it was <2 in the inner layers due to preferential sputtering of oxygen, as already observed for other oxides.[17]

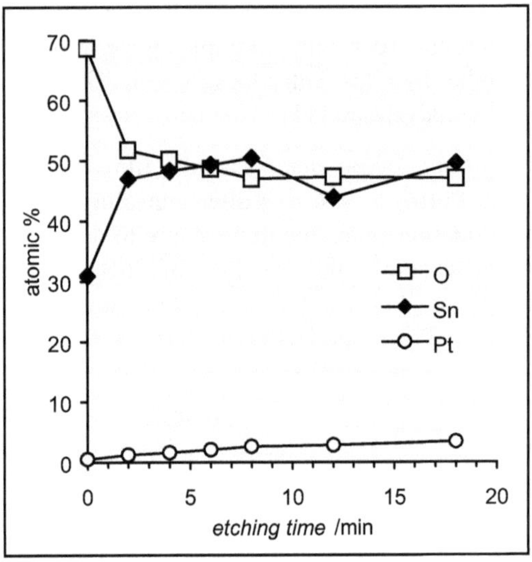

Fig. 5. Elemental composition calculated from in depth-XPS analysis of the Pt-doped SnO_2 film annealed in air at 673 K.

(3) *Possible carbon contamination of the film by organic residues*. XPS in depth-analysis showed also that the C1s line, associated to organic contaminants, was only detected in the outer film layers and completely disappeared after a mild sputtering cycle, thus suggesting that carbon can be merely associated to adventitious surface contamination. The findings suggested that the adopted thermal annealing in air was effective for the complete transformation of the precursors into the desired oxide matrix.

Electrical sensitivity: In order to check the sensing properties of the films, electrical measurements were performed on double-layered films deposited on Suprasil quartz slides (10×10 mm, 0.25 mm thick). Suprasil is a pure glass free from ions affecting the electrical measurements. Two gold films (10×4mm) were deposited at a distance of 2 mm from each other on thin films by d.c. sputtering. Such a procedure was carried out on films annealed at 673 and 973 K in synthetic air (10 $l \cdot h^{-1}$) for 2 h, before any gas treatment. The samples were put in a quartz chamber, placed in an oven, in order to test the gas sensing properties in the temperature range 373–623 K. The electrical resistance was measured by a Keithley 617 programmable electrometer using the constant current method.

The sensing element was equilibrated in flowing air ($10\ l \cdot h^{-1}$), then a CO (800 ppm)/air mixture was introduced ($10\ l \cdot h^{-1}$) and the resistance recorded up to equilibrium conditions. The starting conditions of the film were restored by air equilibration, before introducing again the CO/air mixture. Such a procedure was repeated several times. Sensing properties were evaluated by comparing the sensitivity S of the different samples.

The variation of sensitivity S as a function of temperature for the two differently air-annealed (673 and 973 K) Pt-doped SnO_2 films is shown in (Fig. 6). Sensitivity of Pt-doped SnO_2 films was higher than for pure SnO_2 (not reported in Fig. 6), which showed values lower than 2 in the studied temperature range. In particular, the sensitivity of the Pt-doped SnO_2 film annealed at 673 K showed a constant increase with temperature and reached, at 623 K, a value about 9 times greater than that of the film annealed at 973 K, which revealed a maximum at 623 K.

This sensitivity variation as a function of annealing conditions is mainly attributed to the increase of the mean particle size from 3 to 6 nm for thermal treatment at 673 and 973 K, respectively.

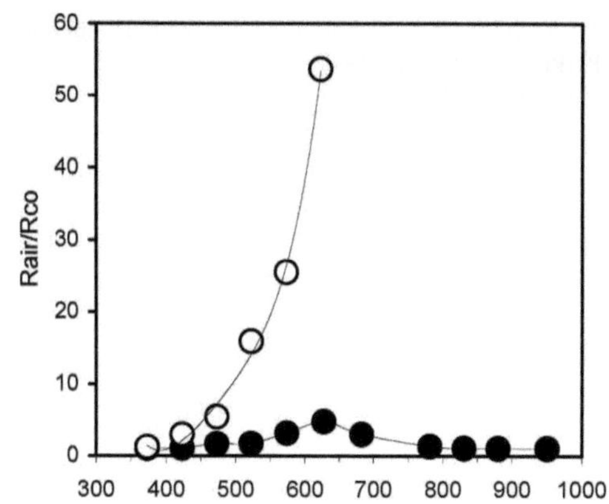

Fig. 6. Dependence of the electrical sensitivity S (R_{air}/R_{CO}) on the temperature for Pt-doped SnO_2 films in CO(800 ppm)/air flow. Open and full circles indicate films annealed in air at 673 K and 973 K, respectively.

References

[1] M. Schweizer-Berberich, J. G. Zheng, U. Weimar, W. Göpel, N. Bârsan, E. Pentia, A. Tomescu, Sensors, Actuators B 1996, **31**, 71.

[2] O. K. Varghese, L. K. Mahlotra, G. L. Sharma, Sensors Actuators B 1999, **55**, 161.

[3] N. Yamazoe, N. Miura in Chemical Sensor Technology, vol. 4, Ed. S. Yamauchi, Elsevier, New York , p. 19, 1992.
[4] J. G. Duh, J. W. Jou, B. S. Chiou, J. Electrochem. Soc. 1989**, 136**, 2740
[5] S. Matsushima, Y. Teraoka, N. Miura, N. Yamazoe, Jp. J. Appl. Phys. 1989**, 27**, 1798.
[6] C. Canevali, N. Chiodini, F. Morazzoni, R. Scotti, C. L. Bianchi, Int. J. Inorg. Mat. 2000, **2**, 355.
[7] C. Canevali, N. Chiodini, F. Morazzoni, R. Scotti, J. Mater. Chem. 2000, **10**, 773.
[8] R. Rella, A. Serra, P. Siciliano, L. Vasanelli, G. De, A. Licciulli, Thin Solid Films, 1997, **304**, 339.
[9] C. Savianu, A. Arnautu, C. Cobianu, G. Craciun, C. Fueraru, M. Zaharescu, C. Parlog, F. Paszti, A. van den Berg, Thin Solid Films 1999**, 349**, 29.
[10] C. J. Brinker, G. W. Scherer, Sol-Gel Science: The Physics, Chemistry of Sol-Gel Processing, Academic Press, New York, 1990.
[11] F. Morazzoni, C. Canevali, N. Chiodini, C. Mari, R. Ruffo, R. Scotti, L. Armelao, E. Tondello, L. E. Depero, E. Bontempi, Chem. Mater. 2001, **13**, 4355.
[12] C. Canevali, N. Chiodini, C. M. Mari, F. Morazzoni, R. Ruffo, R. Scotti, L. Armelao, E. Tondello, Proc. 5th Ital. Conf. Sens. Microsyst., Word Scientific Publ., Singapore, 2001, 186.
[13] F. Morazzoni, C. Canevali, N. Chiodini, C. M. Mari, R. Ruffo, R. Scotti, L. Armelao, E. Tondello, Mat. Sci. Eng. C 2001**, 15**, 167 .
[14] L. Armelao, D. Barreca, E. Bontempi, C. Canevali, L. E. Depero, C. Mari, R. Ruffo, R. Scotti, E. Tondello, F. Morazzoni, Appl. Magn. Reson. 2002, **22**, 89.
[15] M. J. Hampden-Smith, T. A. Wark, A. Rheingold, J. C. Huffman, Can. J. Chem. 1991, **69**, 121.
[16] J. F. Moulder, W. F. Stikle, P. E. Sobol, K. D. Bomben, Handbook of X-ray Photoelectron Spectroscopy, Perkin-Elmer Corp., 1992.
[17] R. Kelly, Surf. Sci. 1980, **100**, 85.

Organically Functionalized Silica Nanoparticles

G. Kickelbick, D. Holzinger and S. Ivanovici

Abstract Dense silica nanoparticles are prepared by hydrolysis and condensation of tetraethoxysilane in methanol (Stöber method). The thus prepared colloids are surface-functionalized by reaction with various functionalized alkoxysilanes.

Classification

form:	nanoparticles
function:	nanocomposites, filler for coatings
preparation:	sol-gel processing, surface chemistry
composition:	organically functionalized SiO_2

Introduction

Nanoparticles have a variety of potential applications because of their specific properties derived from the nanometer length scale.[1] Metal and semiconductor nanoparticles reveal extraordinary properties due to their particular electronic structure between bulk and molecular materials.[2,3] However, for many applications the particles have to be embedded into matrices, for example organic polymers, forming so called nanocomposites. One major problem arising by the incorporation of nanoparticles into various organic matrices is their agglomeration due to their high surface energy and the incompatibility between the surface and the matrix. As a result of this aggregation the nanoparticles loose their specific individual properties. In addition to the above mentioned special electronic behavior, there is another property which makes these particles interesting for various applications. Nanoparticles with diameters much smaller than the wavelengths of visible light do not scatter light and thus transparent materials can be generated, if, for example the particles are included in polymers. Of course, this also depends on other properties, such as differences in the refractive index between nanoparticle and matrix, but generally optical transparent materials can be obtained if the particles are small enough. An important topic with respect to an incorporation of inorganic nanoparticles into a polymer matrix is surface-functionalization to overcome phenomena like aggregation, phase separation, or instabilities in the final materials.[4,5]

The described procedure allows the preparation and surface-functionalization of silica nanoparticles with covalently attached organic groups. The silica nanoparticles are formed by the so called Stöber process.[6] This process is based on the hydrolysis and condensation of tetraethoxysilane (TEOS) in the presence of ammonia in methanol. It produces nearly monodisperse non-porous silica nanoparticles with diameters between a few and several hundred nanometers depending on the reaction conditions.

Various methods can be applied for their surface functionalization. The covalent attachment of silane coupling agents are probably the most prominent.[7,8] The availability of a broad variety of different functional groups opens the possibility of tailoring the surface properties of the particles over a large range. The covalent attachment to the surface of the silica nanoparticles occurs by condensation reactions between the silanes and surface OH-groups. Mono- and trialkoxysilanes $R_{3-n}Si(OR'')_n$ or their chlorine derivatives $R_{3-n}SiCl_n$ are most often used for this purpose.

In this contribution a general procedure for the preparation of 5 nm silica nanoparticles by the Stöber process is described. Four different functional groups were attached to the surface of these particles, by means of 3-methacryloxypropyltrimethoxysilane (MEMO), (3-glycidoxypropyl)-trimethoxysilane (GLYMO), 2-[4-(chloromethyl)phenyl]ethyltriethoxysilane and hexadecyltrimethoxysilane.

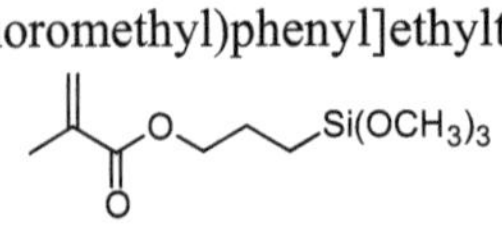

3-methacryloxypropyltrimethoxysilane (MEMO)

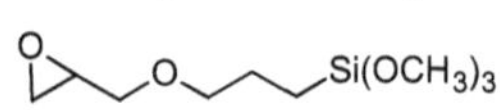

(3-glycidoxy-propyl)-trimethoxysilane (GLYMO)

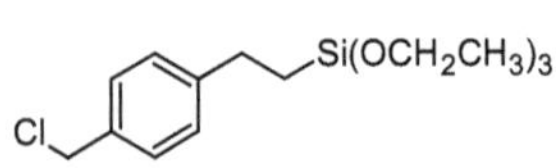

2-[4-(chloromethyl)phenyl]ethyltriethoxysilane

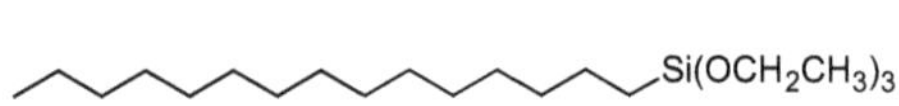

hexadecyltriethoxysilane

MEMO and GLYMO contain functional groups that allow using the surface-modified nanoparticles as comonomers in methacrylate or epoxide polymerizations. 2-[4-(Chloromethyl)phenyl]ethyltriethoxysilane is a potential initiator for atom transfer radical polymerization that permits a grafting from polymerization from the surface of the particles.[9,10] In contrast, hexadecyltrimethoxysilane cannot lead to a covalent linkage to polymers but offers the possibility to prepare suspensions with non-polar solvents, monomers or polymers.

Materials

- Tetraethoxysilane (TEOS) purchased from Fluka, purity >98 %, used as received.
- 3-Methacryloxypropyltrimethoxysilane (MEMO) purchased from Fluka, purity >98%, used as received.
- (3-Glycidoxy-propyl)-trimethoxysilane (GLYMO) purchased from ABCR, purity 98 %, used as received.
- 4-Vinylbenzyl chloride purchased from Aldrich, purity 90 %, used as received.
- Karstedt catalyst purchased from Aldrich, 0.1 M in xylene, used as received.
- Hexadecyltrimethoxysilane purchased from ABCR, used as received.
- Methanol purchased from LOBA Chemie, purity 99.8 %, used as received.
- Triethoxysilane purchased from Wacker, purity >99 %, used as received.
- Toluene purchased from Aldrich, refluxed over CaH_2 and distilled in an argon atmosphere.

Safety and Disposal

Safety and handling instructions for the chemicals are found in the corresponding materials safety data sheets (MSDS). The employed silanes may react vigorously with water. The described experiments should only be carried out in a well equipped chemical laboratory in a fume hood.

Procedures

A. Synthesis of SiO_2 Nanoparticles

Methanol (100 mL) was mixed with 51 mg (1.0 mmol) of 33 % ammonia and 1.98 g (110 mmol) of water and stirred for 5 min. Then 10.41 g (50 mmol) TEOS were added dropwise. The solution was stirred for three days and the particles were isolated via centrifugation or by precipitation with a non-solvent (hexane). The resulting SiO_2 nanoparticles were isolated in vacuo and washed with ethanol and water several times. Yield: 3.7 g colorless powder.

Diameter: 5 nm ± 1 nm (TEM), 5.4 ± 1.3 nm (DLS). DLS measurements were performed applying a non-invasive backscattering technique. The determination of the particle diameter was performed via distribution function and cumulant analysis using the $g_2(t)$ method by a number weighted approach. A 90° angle was used for the measurements. All measurements were carried out in ethanol.

BET-surface (Prior to each nitrogen sorption measurement the samples were degassed at 60°C for at least 4h to a remaining pressure lower than 10 μbar): 529 ± 4 m^2/g

^{13}C CPMAS NMR: 48 (**CH_2OH**), 40 (**CH_2OSi**), 7 (**CH_3**) ppm.

^{29}Si CPMAS NMR: -105 / -116 / -121 ppm. (Q-units)

TGA (heating rate of 5°C/min, in air): < 220 °C: 8.03 %; 220-800°C: 5.33 %

Elemental analysis: C. 2.37, H. 1.31 (caused by residual OEt groups).

B. Preparation of 2-[4-(Chloromethyl)phenyl]ethyltriethoxysilane

4-Vinylbenzyl chloride (6.00 g, 39.3 mmol) was diluted in 10 ml toluene under an argon atmosphere. Three drops of a Karstedt catalyst solution and 9.68 g (58.9 mmol) of triethoxysilane were added, and the mixture was heated under reflux for 48 h. After evaporation of the toluene and the excess of silane in vacuo, 9.45 g (87.2 %) yellow oil were obtained.

Elemental analysis: calc. 56.9 C, 7.8 H; found 56.8 C, 7.9 H.

^{1}H NMR (δ, $CDCl_3$): 7.3-7.2 (m, 5H, phenyl), 4.44 (s, 2H, Cl**CH_2**), 3.77 (q, 6H, SiO**CH_2**), 2.66 (t, 2H, SiCH_2**CH_2**-phenyl), 1.14 (m, 9H, SiOCH_2**CH_3**), 0.91 (t, 2H, Si**CH_2**CH_2-phenyl) ppm.

^{13}C NMR (δ, $CDCl_3$): 140, 135, 127.6-127.4 (phenyl), 58.5 (SiOCH_2), 50.2 (ClCH_2), 27.7 (phenyl-CH_2CH_2Si), 20.4 (phenyl-CH_2CH_2Si), 17.3 (SiOCH_2CH_3) ppm.

^{29}Si NMR (δ, $CDCl_3$): -46.8 ppm.

C. Functionalization of the Silica Particles

a) Functionalization at room temperature in methanol

The particles were not isolated before the functionalization process. Half of the obtained dispersion in methanol (approximately 1.85 g of silica particles) was degassed in vacuo for several minutes to eliminate excessive ammonia. Afterwards, the silane was added in high excess and stirred for three days at room temperature. The functionalized particles were precipitated by adding 100 mL of absolute toluene and isolated via centrifugation at 6000 RPM. Then the particles were washed several times with toluene and ethanol to remove an excess of the silane. Analytical data are given in Table 1.

b) Functionalization under reflux in toluene

The particles were isolated and dried before the functionalization process. An amount of 1.2 g of the particles was dispersed in 20 mL of absolute toluene under an argon atmosphere. The silane was added dropwise in large excess. The obtained mixture was heated under reflux over night. Afterwards, the particles were isolated via centrifugation at 6000 RPM and washed several times with ethanol and toluene to remove adsorbed silane. Analytical data are given in Table 2.

Characterization

Results from the characterization of the obtained purified surface-functionalized particles are given in Tables 1 and 2. ^{13}C and ^{29}Si CPMAS NMR data reveal that the attachment of the organic groups to the surface of the particles was successful. Elemental analysis confirmed that the surface functionalization with the organic molecules was high for both synthesis pathways (reflux or at room temperature). These results were also underpinned by the results obtained from dynamic light scattering that show an increase in the particle diameter.

Table 1. Analytical data of room temperature-functionalized silica particles (method a)

	3-Methacryloxy-propyltrimethoxysilane	(3-Glycidoxypropyl)-trimethoxysilane
Amount silane	6.21 g (0.025 mol)	5.90 g (0.025 mol)
Yield	1.10 g (59.6 %)	1.09 g (58.9 %)
Diameter (TEM[a]/DLS[b])	6 nm / 7.2 ± 1.4 nm	8 nm / 9.2 ± 0.8 nm
BET surface	275 ± 3 m^2/g	291 ± 3 m^2/g
^{13}C CPMAS NMR	158 (COO), 125 ((OOC)C), 112 (CCH_2), 56 (OCH_2), 45 (OCH_3), 11 (CCH_3), 8 ($CH_2CH_2CH_2$), -3 ($SiCH_2$) ppm.	63 ($OCH_{2\ oxiran}$), 61 (CH_2CH_2O), 49 (CH_{oxiran}), 39 (OCH_3), 32 ($CH_{2\ oxiran}$), 12 ($CH_2CH_2CH_2$), 8 (CH_3CH_2O, by-product), -3 ($SiCH_2$) ppm.
^{29}Si CPMAS NMR	-61 / -69 (T-units), -105 / -112 / -123 (Q-units) ppm.	-60 / -68 (T-units), -103 / -112 / -122 (Q-units) ppm.
TGA (mass loss)	< 220 °C: 9.4 % 220-800 °C: 21.8 %	< 220 °C: 17.2 % 220-800°C: 18.5%
Elemental analysis	C: 17.98; H: 3.08.	C: 17.24; H: 3.25.
Surface functionalization (mol funct./g funct. particle)	$1.26 \cdot 10^{-3}$	$1.51 \cdot 10^{-3}$

[a] The samples were prepared via evaporation of the solvent of the particle suspension on a TEM Formvar grid. [b] Monomodal size distribution.

Table 2. Analytical data of silica particles functionalized under reflux (method b)

	Hexadecyl-trimethoxysilane	2-[4-(chloromethyl)phenyl]-ethyltriethoxysilane
Amount silane	3.6 g ($1.1\cdot10^{-2}$ mol)	3.5 g ($1.1\cdot10^{-2}$ mol)
Yield	0.92 g (76.9 %)	0.86 g (71.8 %)
Diameter (TEM[a]/DLS[b])	5 nm / 7.0 ± 1.1 nm	7 nm / 6.8 ± 1.2 nm
BET surface	362 ± 3 m^2/g	416 ± 2 m^2/g
^{13}C CPMAS NMR	40 (OCH_3), 25-15 (CH_2), 11 (CH_2CH_3), 4 ($SiCH_2$) ppm.	135 (CH_2-Cl), 118 (CH_{phenyl}), 115 (phenyl-CH_2CH_2), 50 (OCH_2Cl), 41 (OCH_3), 19 (phenyl-CH_2CH_2), 7 ($SiCH_2$) ppm.
^{29}Si CPMAS NMR	-60 / -65 (T-units), -105 / -111 / -122 (Q-units) ppm.	-65 (T-unit), -102 / -111 /-120 (Q-units) ppm.
TGA (mass loss)	< 220 °C: 9.8 % 220–800 °C: 19.5%	< 220 °C: 2.2 % 220–800 °C: 27.2 %
Elemental analysis	C: 20.21; H: 3.94	C: 18.65; H: 2.66
Surface functionalization (mol funct./g funct. particle)	$0.8\cdot10^{-3}$	$0.93\cdot10^{-3}$

[a] The samples were prepared via evaporation of the solvent of the particle suspension on a TEM Formvar grid. [b] Monomodal size distribution.

TEM images of the resulting particles usually show agglomeration in the dried state (Fig. 1). This is due to the conditions of TEM where the particles are suspended in a suitable solvent. This suspension is dropped on a carbon grid and the solvent is evaporated. During this process the particles usually agglomerate due to van-der-Waals forces between the particles. However in suspension the particles are homogeneously distributed and only small agglomeration occurs, which is revealed by the asymmetry of the signals in the DLS measurements.

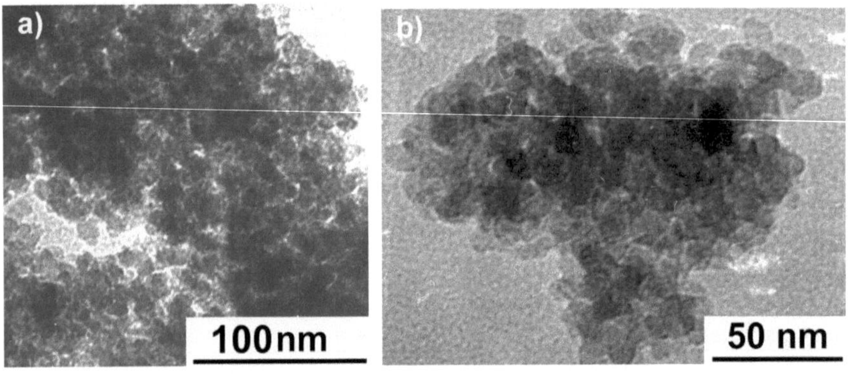

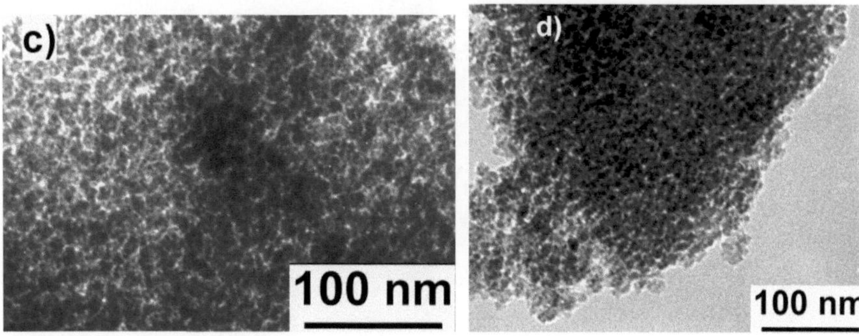

Fig. 1. TEM images of SiO_2 nanoparticles a) functionalized with 3-methacryloxy-propyltrimethoxysilane, b) functionalized with (3-glycidoxypropyl)trimethoxysilane, c) functionalized with hexadecyltrimethoxysilane, and d) functionalized with 2-[4-(chloromethyl)phenyl]ethyltriethoxysilane.

References

[1] G. Schmid (Ed.), Nanoparticles, Wiley-VCH, Weinheim, Germany 2004.
[2] H. Weller, Curr. Opinion Colloid Interface Sci. 1998, **3**, 194-199.
[3] T. Trindade, P. O'Brien, N. L. Pickett, Chem. Mater. 2001, **13**, 3843-3858.
[4] G. Kickelbick, Progr. Polym. Sci. 2002, **28**, 83-114.
[5] G. Kickelbick, U. Schubert, in M.-I. Baraton (Ed.): Synthesis, Functionalization and Surface Treatment of Nanoparticles, American Scientific Publishers, Stevenson Ranch, CA, USA, 2003, p. 91-102.
[6] W. Stöber, A. Fink, E. Bohn, J. Colloid Interface Sci. 1968, **26**, 62-69.
[7] A. Van Blaaderen, A. Vrij, J. Colloid Interface Sci. 1993, **156**, 1-18.
[8] C. Beck, W. Härtl, R. Hempelmann, Angew. Chem., Int. Ed. 1999, **38**, 1297-1300.
[9] D. Holzinger, G. Kickelbick, Chem. Mater. 2003, **15**, 4944-4948.
[10] D. Holzinger, G. Kickelbick, J. Mater. Chem. 2004, **14**, 2017-2023.

Copper Nanoparticles in Silica

U. Schubert, C. Lembacher and G. Trimmel

Abstract Copper nanoparticles in a silica matrix are prepared by a three-step procedure. In the first step a copper salt is reacted with an alkoxysilane of the type $(RO)_3Si(CH_2)_nA$, where A is a coordinating organic group. The obtained metal complexes $\{[(RO)_3Si(CH_2)_nA]_xCu]^{2+}$ are used as precursors for sol-gel processing, with $Si(OR)_4$ as co-reactant to adjust the metal:silica ratio. In the second step, the metal complex-containing gels are calcined in air at high temperatures, and metal oxide nanoparticles in a silica matrix are formed. Finally, the metal oxide nanoparticles are reduced to metal nanoparticles.

Classification

form:	composite powder
function:	catalytic material, dielectric material
preparation:	sol-gel processing
composition :	Cu / SiO_2

Introduction

Metal nanoparticles in a ceramic matrix are interesting dielectric materials, due to quantum size effects in the metallic particles, or for applications as heterogeneous catalysts. For the latter application, the matrix must be sufficiently porous.

Sol-gel processing allows the preparation of metal or metal oxide nanoparticles in a silica matrix with adjustable metal loadings, if the metal precursors are dispersed on an atomic level during sol-gel processing and metal oxide or metal particles are then grown by controlled thermal treatment and reduction. The high dispersion is achieved by employing complexing alkoxysilanes to coordinate metal ions during sol-gel processing.

Metal-silica nanocomposites are prepared in a three-step procedure.[1-4] In the first step, a metal salt MX_z is reacted with a silane of the type $(RO)_3Si(CH_2)_nA$, where A is an organic group capable of coordinating metal ions. Metal complexes

$\{[(RO)_3Si(CH_2)_nA]_xM\}^{z+}$ are formed, which do not need to be isolated. Upon sol-gel processing of the alkoxysilyl-substituted metal complexes, the metal coordination is retained, and the metal complexes are tethered to the silicate matrix via the $(CH_2)_nSiO_{3/2}$ groups. The metal loading can be adjusted by adding $Si(OR)_4$ to the precursor solution. The resulting gels have the idealized composition $[O_{3/2}Si(CH_2)_nA]_nMX_z{\cdot}xSiO_2$ and the typical color of the corresponding metal complexes.

In the second step, the metal complex-containing gels are calcined in air at high temperatures. The tethering organic groups are thus destroyed. Due to the high dispersion of the metal ions in the first step, nanosized metal oxide particles (i.e., nanocomposites $MO_y{\cdot}(x+n)SiO_2$) are formed. The obtained powders have the color of the corresponding metal oxide. If one wants to get carbon-free composites, the oxidation temperature has to be high enough to ensure complete oxidation of all organic components, but should not be higher than necessary to avoid excessive sintering of the metal particles. The metal oxide nanoparticles are reduced to metal nanoparticles in the third step, by which composites $M{\cdot}(x+n)SiO_2$) are obtained.

The metal oxide or metal nanoparticles are highly dispersed in the SiO_2 matrix and not agglomerated, even in materials with high metal loadings. The particle diameters are typically in the range 2–25 nm. The size distributions are very narrow as determined by TEM investigations. The metal particles are accessible, because of the porosity of the silica gel matrix. This is an important issue if the composites are used as heterogeneous catalysts.

The metal particle size is influenced by the complexing silane/metal ratio, the calcination conditions, the counter ion of the employed metal salt, the kind of the complexing silane and the method of the way how the organic groups are removed.[2,3] Deliberate change of these parameters allows varying the metal particle diameters of the composites.

While the metal complex-containing gels are essentially non-porous, pyrolysis / thermolysis of the organic groups creates micropores. The metal oxide-silica and metal-silica composite powders thus have relatively large surface areas. The size and shape of the pores, and thus the specific surface area, is influenced by the kind of organic groups (counter-ion of the metal salt and organic groups at the complexing silane).[4]

Highly dispersed metals on solid supports can also be prepared by this method. For this application, the metal complex-containing sols obtained after the first step are sprayed onto the supports and then converted into metal particles in silica.[3] Metal-doped aerogels were obtained by supercritical drying of the metal complex-containing gels.[5,6]

This contribution describes the preparation of $Cu{\cdot}4SiO_2$ by using $(RO)_3Si-(CH_2)_3NHCH_2CH_2NH_2$ as the complexing silane. Scheme 1 shows a flow chart of the synthesis procedure. Cu/SiO_2 nanocomposites with a different copper loading are prepared in the same way by varying the $Si(OEt)_4$ proportions in the starting mixture. Other metal / silica composites can be prepared by the same protocol, for example, Ag, Co, Ni, Pd, Pt or mixed-metal particles in silica.[1-6]

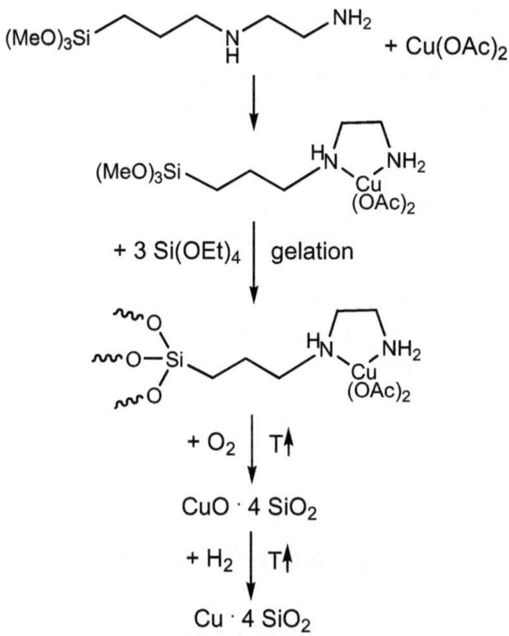

Scheme 1. Flow chart of synthesis protocol

Materials

- Copper acetate monohydrate, $Cu(OAc)_2 \cdot H_2O$, purity >99 %, supplied by Mallinkrodt Chemical Works, used as received.
- [N-(aminoethyl)aminopropyl]trimethoxysilane, $H_2NCH_2CH_2NH(CH_2)_3$-$Si(OMe)_3$, purity >99 %, supplied by Wacker AG, used as received.
- Tetraethoxysilane, $Si(OC_2H_5)_4$, purity >98 %, supplied by Merck, used as received.
- Ethanol, purity 99.8%, supplied by Austria Hefe AG, used as received.
- 0.2 n NH_4OH solution, prepared by diluting 5.00 n NH_4OH volumetric standard (supplied by Aldrich) with distilled water.
- Hydrogen, purity 99.999 %, supplied by Messer Griesheim, used as received.

Safety and Disposal

Safety and handling instructions for the chemicals are found in the corresponding materials safety data sheets (MSDS). Protective gloves and safety glasses should be worn during all operations.

Tetraethoxysilane is flammable (b.p. 168°C) and may vigorously react with water (which gives silica and ethanol). It is an eye and lung irritant, and may damage the liver and kidneys. [N-(aminoethyl)aminopropyl]trimethoxysilane may also react vigorously with water. There is the danger of severe eye damage.

Hydrogen / air mixtures are highly explosive. Therefore, the equipment used for hydrogen reductions has to be carefully flushed with an inert gas before heating the sample in hydrogen.

Procedure

A. Sol-Gel Processing of Alkoxysilyl-substituted Metal Complexes

All operations are carried out in a 100 ml flask exposed to air. An amount of 1.8 g (8.0 mmol) of [N-(aminoethyl)aminopropyl]trimethoxysilane is slowly added to a suspension of 1.6 g (8.0 mmol) of copper acetate monohydrate in 50 ml of ethanol. The mixture is stirred at 22°C for about 30 min, until the copper salt is completely dissolved. The color of the solution changes from blue to dark blue. Then 16.2 mL of a 0.2 n aqueous ammonia solution is added. This corresponds to 7.5 molar equivalents of water per Si-OR group (OR groups from both $R'Si(OMe)_3$ *and* $Si(OEt)_4$). The solution is stirred for 15 min at 22°C, and then 5.0 g (24 mmol) of tetraethoxysilane is added. When the solution is refluxed at 70°C, gelation occurs after about 8 h. Heating is continued for additional 64 h. Then all volatiles are removed at 40 Torr, and the remaining solid is dried at 70°C / 0.1 Torr until weight constancy. Yield: 4.25 g of a dark blue amorphous powder (theor. 4.12 g).

Characterization

Elemental analysis: Calcd. for $[Cu(OAc)_2(H_2NCH_2CH_2NH(CH_2)_3SiO_{3/2}) \cdot 3\ SiO_2]$: C, 20.98; H, 3.72; N, 5.44. Found: C, 19.51; H, 3.83; N, 5.63.

UV spectrum (Perkin Elmer Lambda 15 with integrating sphere attachment; solid samples were diluted with $BaSO_4$): $\lambda_{max} = 702$ nm.

Specific surface area (determined by the BET method [N_2 adsorption], Micromeritics ASAP 2010) $11 \pm 2\ m^2 \cdot g^{-1}$. The sample was dried at 80°C / 7 mTorr until weight constancy prior to the measurements.

DSC (heating rate 5 $°min^{-1}$ in air): one exothermic peak at 303°C.

TGA (heating rate 5°min^{-1} in air): continuous weight loss between 80°C and about 450°C. Total weight difference 38.4% (calcd. for conversion to $CuO \cdot 4\ SiO_2$: 37.8 %).

XRD: Only a broad band at $2\theta = 22.19°$ (amorphous silica).

Comments

(1) During this step, the metal complex

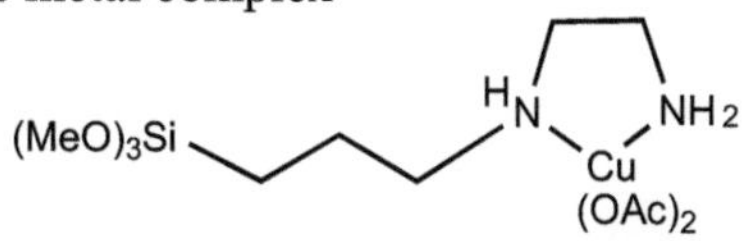

is initially formed *in situ*. Its UV spectrum (λ_{max} = 661 nm) corresponds to that of the corresponding ethylene diamine (en) complex $Cu(en)Cl_2$ in aqueous solution,[7] i.e. the presence of the $(CH_2)_3Si(OMe)_3$ side chain has no influence on the general composition of the copper complex. If another complexing group or a different copper / [N-(aminoethyl)aminopropyl]trimethoxysilane ratio is chosen, different metal complexes will be formed.

(2) The UV spectrum after gelation is very similar and shows that the complex is retained upon sol-gel processing. The wavelength shift is explained by the different environment of the complex (gel matrix instead of the solvent).

(3) The differences between the calculated and found values in the elemental analysis are explained by the presence of residual (non-hydrolyzed) Si-OR groups and / or the presence of adsorbed water and / or Si-OH groups. These groups are removed during the thermal treatment in the second step of the synthesis.

B. Oxidation

The powder obtained in the first step is ground in a ball mill (Retsch MM2, 40 Watt, 15 min, agate balls). The fine powder is transferred to a glazed ceramic boat which is placed in a conventional oven. The oven is heated to 550°C in air with a heating rate of 10°min^{-1} and then kept at 550°C for 1 h. The total weight loss by the thermal treatment is 38.5%. A green powder is obtained.

Characterization

Elemental analysis: Calcd. for $CuO \cdot 4\ SiO_2$: C, 0.0; H, 0.0; N, 0.0. Found: C, 0.14; H, 0.85; N, <0.05.

Specific surface area: 477±7 $m^2 \cdot g^{-1}$. The sample was dried at 250°C / 7 mTorr until weight constancy prior to the measurements. Average pore diameter: 2 nm (from BET analysis).

XRD (Phillips PW 1710 in Bragg Brentano geometry; Cu-K_α radiation; 2θ range 7–70°): Broad band at 2θ ~22.19° (amorphous silica); d (2θ) = 2.777 (32.24), 2.525 (35.56), 2.330 (38.64), 2.316 (38.88) 1.866 (48.81), 1.505 (61.62), 1.410 (66.25) (tenorite, CuO, JCPDS entry number 45-0937: d = 2.75300, 2.52700, 2.32300, 2.31000, 1.86730, 1.50580, 1.40960). Average CuO particle diameter (determined from the width of the superimposed 002 and 111 reflection (d = 2.525)) 15.5 nm.

Comments

(1) The organic groups are removed during this step, and CuO nanoparticles are formed. If one wants to get carbon-free composites, the oxidation temperature has to be high enough to ensure complete oxidation of all organic components, but should not be higher than necessary to avoid excessive sintering of the metal particles. The maximum temperature necessary for the removal of all organic groups and the holding time can be optimized by TGA. Alternative methods for the removal of the organic groups include a combination of pyrolysis and calcination in air or treatment by oxygen plasma.

(2) Particle size determination from the line broadening of XRD reflections is a very convenient method to determine the *average* particle size. It should be kept in mind, however, that only the *crystalline* proportion of particles, with diameters of >2–50 nm, can be determined. Particle size *distributions* can be determined by electron microscopy.

C. Reduction

The powder obtained in the oxidation step is transferred to ceramic boat which is placed a horizontal quartz tube (2 cm diameter, 60 cm heated length) positioned in a tube furnace. The tube is flushed with hydrogen, then heated to 400°C with a heating rate of $10° \cdot min^{-1}$. The temperature is then kept at 400°C for 1 h. During heating, a steam of hydrogen (200 $ml \cdot min^{-1}$) is passed over the sample. The furnace is then allowed to cool while the tube is flushed with argon. The total weight loss in this step is 2 %. A gray powder is obtained.

Characterization

Elemental analysis: Calcd. for Cu · 4 SiO_2: C, 0.0; H, 0.0; N, 0.0. Found: C, 0.14, H, 0.69, N, 0.05.

Specific surface area: 421±7 $m^2 \cdot g^{-1}$. The sample was dried at 250°C / 7 mTorr until weight constancy prior to the measurements. Average pore diameter: 2 nm (from BET analysis).

XRD: Broad band at 2θ ~ 22.19° (amorphous silica); d(2θ) = 2.086 (43.38), 1.807 (50.51) (elemental Cu, JCPDS entry number 04-0836: d = 2.08800, 1.80800). Average Cu particle diameter (determined from the width of the 111 reflection (d = 2.086) in XRD) 14.9 nm.

Comments

(1) The properties of the final Cu/SiO_2 composite, particularly the copper particle diameter and the specific surface area is to a very high degree determined by chemical parameters during the first step (see Introduction). However, a constant set of parameters gives reproducible results.

(2) The composition of the final composite is determined by the Cu:Si ratio of the starting mixture. Thus, use of a different proportion of $Si(OEt)_4$ results in a different copper loading of the final composite. When no $Si(OEt)_4$ is added, the composition of the composite is $Cu \cdot SiO_2$.

(3) The temperature for reduction does not affect the final particle size if it is lower than the temperature during calcination.

References

[1] B. Breitscheidel, J. Zieder, U. Schubert, Chem. Mater. 1991, **3**, 559. W. Mörke, R. Lamber, U. Schubert, B. Breitscheidel, Chem. Mater. 1994, **6**, 1659.

[2] U. Schubert, C. Görsmann, S. Tewinkel, A. Kaiser, T. Heinrich, Mat. Res. Soc. Symp. Proc. 1994, **351**, 141. A. Kaiser, C. Görsmann, U. Schubert, J. Sol-Gel Sci. Technol. 1997, **8**, 795.

[3] C. Lembacher, U. Schubert, New J. Chem. 1998, **22**, 721. G. Trimmel, C. Lembacher, G. Kickelbick, U. Schubert, New J. Chem. 2002, **26**, 759.

[4] G. Trimmel, U. Schubert, J. Non-Cryst. Solids 2001, **296**, 188.

[5] B. Heinrichs, F. Noville, J.-P. Pirard, J. Catal., 1997, **170**, 366. B. Heinrichs, P. Delhez, J.-P. Schoebrechts, J.-P. Pirard, J. Catal. 1997, **172**, 322.

[6] S. Martínez, M. Moreno-Mañas, A. Vallribera, U. Schubert, A. Roig, E. Molins, New J. Chem. 2006, **30**, 1093.

[7] H. B. Jonassen, T. H. Dexter, J. Am. Chem. Soc. 1949, **71**, 1553.

Copper Nanocrystals

S. D. Bunge and T. J. Boyle

Abstract An anhydrous route for the synthesis of amine capped copper nanoparticles (NP) has been developed using coinage metal mesityl (mesityl = 2,4,6-$Me_3C_6H_2$) derivatives. Under an argon atmosphere, $[Cu(\mu\text{-mesityl})]_5$ was dissolved in octylamine and subsequently injected into a heated hexadecylamine solution (300°C) generating oxide-free spherical copper nanoparticles of 8-9 nm in diameter.

Classification

form:	spherical nanocrystals
function:	catalysts, nano-lubricants
preparation:	thermolysis of organometallic copper precursor
composition:	amine capped Cu

Introduction

Optical properties of coinage-metal nanoparticles (NPs) are inherently dependent on both the particle's size and shape due to particularly strong surface plasmon oscillations within these metals.[1,2] As a result, over the past fifteen years, a considerable amount of interest has been focused on the synthesis, properties, and arrangement of coinage-metal NPs.[3,4] In 1994, Brust *et al.* first reported the two-phase reduction of $AuCl_4^-$ by $NaBH_4$ to afford mono-dispersed 3 nm gold particles passivated by nonanethiol.[4-6] Preparative methods analogous to this route are now widely employed to synthesize monodisperse gold, silver,[7-9] and copper oxide coated copper[10] NPs stabilized by alkanethiols,[3,5,6,11] phosphines,[4] quaternary ammonium salts,[12] surfactants[13] or polymers.[14,15] However, despite numerous reports, only a limited number of synthetic routes are available that require few starting materials, lack reducing agents and/or special additives (i.e., detergents), and ideally produce no salts or other difficult to remove by-products.[3,16-18] In addition, current methods generally require multi-step "seed-growth" processes and post-synthetic purification techniques to generate coinage-metal NPs of various sizes.[19-22]

In contrast to the previously discussed salt reduction method for Group 11 NPs, the synthesis of II-VI semiconductor nanocrystals (i.e., CdSe) typically involves thermolysis of an organometallic cadmium precursor in a coordinating solvent.[23,24] With this synthetic approach, the controlled formation of various sizes and shapes of CdSe relies on the principle of "focused size distribution".[25-28] This principle requires an initial nucleation shower of monodisperse seed particles, followed by a relatively slow and lengthy growth process. Any size variation is compensated by the rapid growth of the relatively small particles in comparison to the larger ones. As a result, the size distribution of the resulting nanocrystals narrows over time.

It has been proposed that the concepts of chemical vapor deposition (CVD) precursor development can be transferred to solution-based approaches for NP syntheses.[16,24,29-33] Typically, molecularly designed metal alkyls, alkoxides, and amides are used as CVD precursors due to their volatility and their clean transport characteristics.[34] It should be noted that the implementation of oligomeric metal complexes as synthons for NPs has been explored for a variety of compositions (i.e., Au and CdSe).[35,36] Typically these complexes are either difficult to synthesize, have undesirable byproducts, or are inflexible oligomeric molecules composed of hundreds of atoms. It is our contention that "small" metal clusters, with easily vaporized byproducts, allow for more control over the growth of the NPs and facilitate the focused sized distribution of the colloids.

Herein, we describe the procedure for using organometallic copper precursors for the production of NPs using the principles of focused size distribution.[37] Crystalline $[Cu(\mu\text{-mesityl})]_5$ was prepared by a procedure described by Floriani *et al.* (Scheme 1).[38] This complex was chosen because of the facile synthesis of the precursor in relatively high yield, the lack of potential halide contaminants, and the fact that this compound is a stable crystalline solid under ambient conditions.

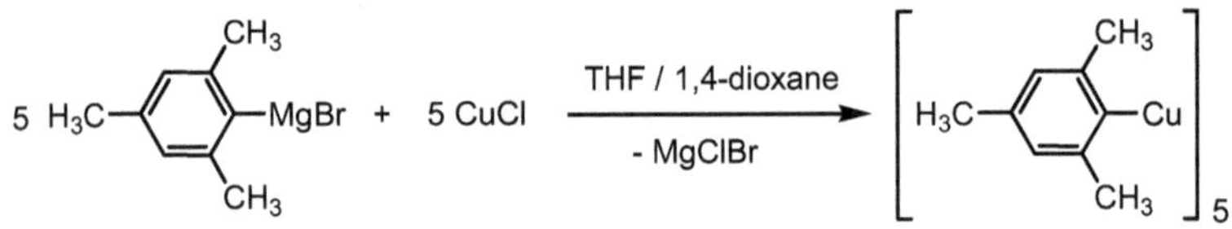

Scheme 1. General synthesis of $[Cu(\mu\text{-mesityl})]_5$.

Materials

- 1-Hexadecylamine purchased from Aldrich Chemical, tech, purity 90 %, used as received.
- Octylamine purchased from Aldrich Chemical, purity 99 %, used as received.

- Toluene, anhydrous, purchased from Aldrich Chemical, purity 99.9%, used as received.
- Methanol, anhydrous, purchased from Aldrich Chemical, purity 99.9%, used as received.
- 1,4-Dioxane, anhydrous, purchased from Aldrich Chemical, purity 99.8%, used as received.
- Copper(I) chloride, purchased from Aldrich Chemical, purity99 %, was freshly re-crystallized from aqueous HCl.
- 2-Mesitylmagnesium bromide, 1.0 M solution in tetrahydrofuran, purchased from Aldrich Chemical, used as received.
- Argon, ultra high purity, Tri-Gas, used as received.

Safety and Disposal

Safety and handling instructions for the chemicals are found in the corresponding materials safety data sheets (MSDS). The described experiments should be carried out in a well ventilated area using either standard Schlenk-line or glovebox techniques. Standard pressure safety precautions are required for operating both glove boxes and Schlenk glassware.

Procedures

On a Schlenk line, under an argon atmosphere, a magnetically stirred hexadecylamine solution (7.0 g, 29 mmol) was heated to 300°C in a 250 mL 3-neck round bottom flask. In a glovebox, a solution of 0.3 M of $[Cu(\mu\text{-mesityl})]_5$ in octylamine (~4 mL) was transferred to a syringe, removed from the glovebox and rapidly injected into the stirring 300°C reaction mixture. Upon injection, the colorless amine solution turned dark red. The solution was heated at 225°C for 30 min, and then cooled to room temperature. Using standard Schlenk techniques, addition of toluene (20 mL) and methanol (100 mL) to the reaction mixture results in a red precipitate and a colorless solution. Under argon, careful removal of the colorless solution and re-dispersion of the precipitate in toluene (20 mL) results in a colloidal solution stable at room temperature in the absence of air.

Alternatively, after cooling to room temperature the reaction flask may be transferred to an argon-filled glove box and particles separated by centrifugation. This is the preferred method since it minimizes oxygen exposure which is critical in obtaining oxide free Cu^0 nanomaterials, as well as allowing for easy manipulation of the desired nanomaterials. In this variation of the procedure, toluene (~20 mL) was added to the solid precipitate (obtained after cooling the reaction flask to room temperature) to yield a deep red dispersion. Further purification was achieved by addition of methanol (~100 mL) forming a clear colorless solution and a red precipitate. The precipitate was separated by centrifugation (3300 rpm/25 min) carried out in an argon-filled glove box and then re-dispersed in toluene (20 mL). The resulting colloid is stable at room temperature in the absence of air.

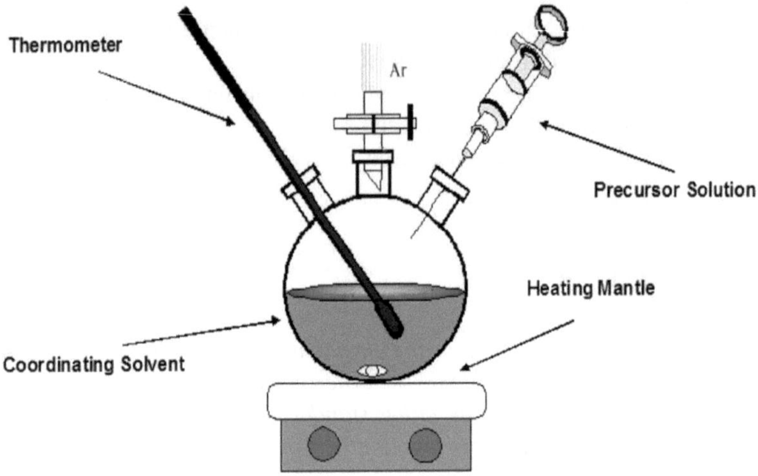

Fig. 1. Schematic diagram for the reaction system for synthesizing copper nanocrystals.

Characterization

The *UV/VIS spectrum* in toluene displays a sharp exciton peak at 568 nm (Fig. 2) consistent with literature reports for particles on the order of 8 nm.[31] *TEM*: In a glove box, under an argon atmosphere, a drop of the Cu° NPs dispersed in toluene was deposited on a 30-mesh carbon coated copper TEM grid. To prevent surface oxidation of the Cu° particles, the TEM sample was then transported to the instrument under an argon atmosphere. The TEM image confirms the formation of Cu° NPs as well-defined, spherical particles with an average diameter of 9.2 ± 2.3 nm (calculated by measuring at least 150 particles) (Fig. 2). The size and shape of the nanocrystals are uniform, and the individual particles are separated by about 2 nm due to shells of hexadecylamine surfactant.[31] The size distribution leads to the formation of hexagonally 2-D ordered lattices of free standing copper colloids.

Fig. 3 displays a HRTEM image of Cu° NPs produced by this method. Notably, without apparent harm to the particles, the anticipated lattice planes for FCC cubic Cu° are readily observed. The corresponding selected area electron diffraction (SAED) pattern was also obtained and is shown in Fig. 3. Devoid of evidence of copper(I) oxide and copper(II) oxide, the four rings correspond to the lattice planes (111), (200), (220) and (311); which is consistent with the face centered cubic phase of copper.

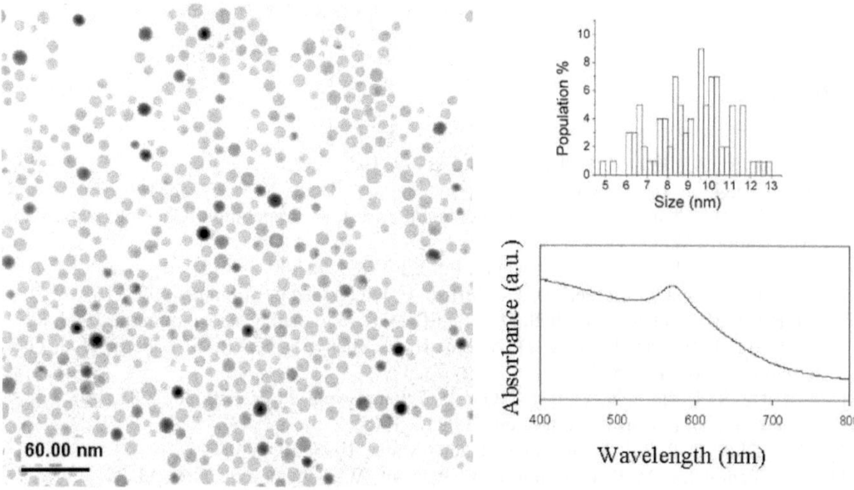

Fig. 2. TEM image, size distribution and UV/VIS absorption spectrum of octylamine capped copper nanoparticles, bar = 60 nm (size distribution is 9.2 ± 2.3 nm).

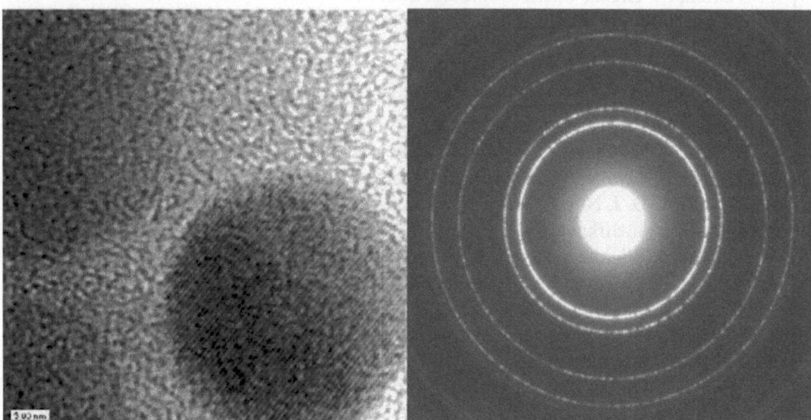

Fig. 3. HRTEM image and selected area electron diffraction pattern of Cu nanoparticles prepared and transported in an argon atmosphere (bar = 3 nm).

Comments

(1) Different precursor concentrations and reaction times result in the formation of copper nanocrystals with various size distributions.

(2) Unless otherwise specified, all syntheses and manipulations were carried out under an atmosphere of argon using standard Schlenk techniques or in an argon-filled glovebox. This is critical for generating oxide-free nanocrystals.

References

[1] S. Link, M. A. ElSayed, J. Phys. Chem. B 1999, **103**, 4212.
[2] J. A. Creighton, D. G. Eadon, J. Chem. Soc. Faraday Trans. 1991, **87**, 3881.
[3] J. H. Fendler, Nanoparticles and Nanostructured Films: Preperation, Characterization and Applications, Wiley-VCH, 1998.
[4] G. Schmid, A. Lehnert, Angew. Chem. Int. Ed. 1989, **28**, 780.
[5] M. Brust, J. Fink, D. Bethell, D. J. Schiffrin, C. Kiely, Chem. Commun. 1995, 1655.
[6] M. Brust, M. Walker, D. Bethell, D. J. Schiffrin, R. Whyman, Chem. Commun. 1994, 801.
[7] S. T. He, J. N. Yao, P. Jiang, D. X. Shi, H. X. Zhang, S. S. Xie, S. J. Pang, H. J. Gao, Langmuir 2001, **17**, 1571.
[8] S. T. He, S. S. Xie, J. N. Yao, H. J. Gao, S. J. Pang, Appl. Phys. Lett. 2002, **81**, 150.
[9] S. H. Chen, D. L. Carroll, Nano Lett. 2002, **2**, 1003.
[10] N. A. Dhas, C. P. Raj, A. Gedanken, Chem. Mater. 1998, **10**, 1446.
[11] A. C. Templeton, M. P. Wuelfing, R. W. Murray, Acc. Chem. Res. 2000, **33**, 27.
[12] J. Fink, C. J. Kiely, D. Bethell, D. J. Schiffrin, Chem. Mater. 1998, **10**, 922.
[13] K. Esumi, N. Sato, K. Torigoe, K. Meguro, J. Colloid Interface Sci. 1992, **149**, 295.
[14] H. Hirai, Y. Nakao, N. Toshima, Chem. Lett. 1976, **1976**, 905.
[15] C. H. Walker, J. V. StJohn, P. Wisian-Neilson, J. Am. Chem. Soc. 2001, **123**, 3846.
[16] P. O'Brien, M. Green, Chem. Commun. 2000, 183.
[17] B. Prasad, S. Stoeva, C. Sorensen, K. Klabunde, Chem. Mater. 2003, **15**, 935.
[18] Y. Zhang, F. Chen, J. Zhuang, Y. Tang, D. Wang, Y. Wang, A. Dong, N. Ren, Chem. Commun. 2002, 2814.
[19] K. V. Sarathy, G. U. Kulkarni, C. N. R. Rao, Chem. Commun. 1997, 537.
[20] N. R. Jana, L. Gearheart, C. J. Murphy, Adv. Mater. 2001, **13**, 1389.
[21] N. R. Jana, L. Gearheart, C. J. Murphy, Chem. Commun. 2001, 617.
[22] N. R. Jana, L. Gearheart, C. J. Murphy, Langmuir 2001, **17**, 6782.
[23] C. B. Murray, D. J. Norris, M. G. Bawendi, J. Am. Chem. Soc. 1993, **115**, 8706.
[24] J. Hambrock, A. Birkner, R. A. Fischer, J. Mater. Chem. 2001, **11**, 3197.
[25] X. G. Peng, J. Wickham, A. P. Alivisatos, J. Am. Chem. Soc. 1998, **120**, 5343.
[26] Z. A. Peng, X. G. Peng, J. Am. Chem. Soc. 2002, **124**, 3343.
[27] M. W. Yu, X. G. Peng, Angew. Chem. Int. Ed. 2002, **41**, 2368.
[28] D. Battaglia, X. G. Peng, Nano Lett. 2002, **2**, 1027.
[29] M. Nakamoto, M. Yamamoto, M. Fukusumi, Chem. Commun. 2002, 1622.
[30] J. J. Wang, L. Grocholl, E. G. Gillan, Nano Lett. 2002, **2**, 899.
[31] J. Hambrock, R. Becker, A. Birkner, J. Weiss, R. A. Fischer, Chem. Commun. 2002, 68.
[32] K. Soulantica, A. Maisonnat, M. C. Fromen, M. J. Casanove, P. Lecante, B. Chaudret, Angew. Chem. Int. Ed. 2001, **40**, 448.
[33] M. Veith, J. Chem. Soc. Dalton Trans. 2002, 2405.
[34] M. J. Hampden-Smith, T. T. Kodas, The Chemistry of Metal CVD, VCH, 1994.
[35] S. L. Cumberland, K. M. Hanif, A. Javier, G. A. Khitrov, G. F. Strouse, S. M. Woessner, C. S. Yun, Chem. Mater. 2002, **14**, 1576.
[36] J. F. Hainfeld, Science 1987, **236**, 450.
[37] S. D. Bunge, T. J. Boyle, T. J. Headley, Nano Lett. 2003, **3**, 901.
[38] E. M. Meyer, S. Gambarotta, C. Floriani, A. Chiesivilla, C. Guastini, Organometallics 1989, **8**, 1067.

Assembly of TOPO-Capped Silver Nanoparticles to Multilayered Films

Z. V. Saponjic, T. Rajh and N. M. Dimitrijevic

Abstract A method for transferring silver nanoparticles from aqueous solution into organic solvents such as toluene or hexane is described. The phase-transfer reagent, tri-n-octylphosphine oxide (TOPO) provides a capping shell around Ag particles that allows concentrating the nanoparticle dispersion in toluene, which is a prerequisite for successful 3D self-assembly. The dispersed particles in toluene are stable for at least six months. Self-assembly of highly concentrated TOPO-capped Ag particle sols results in multilayered mirror-like films on glass substrates.

Classification

form:	film of ~10 nm close-packed silver particles
function:	optoelectronics
preparation:	sol-gel processing, self-assembly
composition:	(n-octyl)$_3$PO modified silver particles

Introduction

The self-assembly of metallic nanoparticles to macroscopic structures offers a pathway for the creation of macrocrystallites with tunable, designer-specified optical, electronic and catalytic properties.[1] Small metal particles and their ensembles exhibit unusual optical and electronic properties that are between bulk and molecules, such as charging steps due to Coulomb blockade,[2,3] thermally activated conductivity by electron hopping,[4-6] or nonlinear optical effects observed in the generation of second-order harmonics.[7]

The described procedure allows the synthesis of multilayered film of well-defined isolated silver particles. The prerequisite for a self-assembly route to macrostructures is (i) availability of stable building blocks of metallic nanoparticles with well-characterized uniform particle sizes and shapes, and (ii) the presence of suitable capping groups (ligands or linkers) that, at the same time, allow interparticle assembly, and preserve domination of repulsive forces between building-block nanoparticles over interparticle irreversible aggregation. The capping groups should provide uniform protection of the surface without modification of the particle's essential structural and electronic properties. Uniform size is necessary for obtaining ordered assemblies. If unprotected colloidal particles are used, the resulting aggregates exhibit properties of the bulk metal. The described procedure uses tri-n-octylphosphine oxide (TOPO) as a capping agent for the formation of stable colloidal silver particles in toluene,[8] and as a separation layer of the nanoparticles in multilayered films. The alkyl chains provide the barrier for particle agglomeration and the optimal capping that accounts for the curvature of spherical particles.

The synthesis procedures involves three steps: (i) synthesis of colloidal Ag particles in aqueous solution, (ii) capping of Ag particles with TOPO, which enables their transfer and concomitant concentration in toluene, and (iii) self-assembly on a glass substrate.

Materials

- Tri-n-octylphosphine oxide (TOPO) purchased from Aldrich, purity 90%, used as received.
- Sodium borohydride, $NaBH_4$, purchased from Aldrich, purity 98%, used as received.
- Magnesium sulfate heptahydrate, $MgSO_4 \cdot 7H_2O$, purity 98%, purchased from Aldrich, used as received.
- Toluene purchased from Aldrich, purity HPLC grade 99.8%, used as received.
- Silver nitrate, $AgNO_3$, purchased from B&A, reagent grade purity, used as received.
- Argon gas, high purity 99,998 purchased from AGA Gas.

Safety and Disposal

Safety and handling instructions for the chemicals are found in corresponding material safety data sheets (MSDS).

Procedures

A. Preparation of Silver Hydrosols

The preparation of silver hydrosols is based on the reduction of silver cations as described in the literature.[9] A 100 ml portion of a $2 \cdot 10^{-4}$ M aqueous solution of $AgNO_3$ (3.4 mg) is placed in a 600 ml flask. Ar was bubbled through the solution for 15 minutes. An amount of 10 mg of $NaBH_4$ was then added under vigorous stirring (magnetic stirrer) and continuous Ar bubbling. The resulting transparent, yellow silver hydrosol was stored in the absence of light under Ar atmosphere.

Characterization

UV/Vis spectra of the silver particle sol exhibit characteristic plasmon absorption with a maximum around 390 nm and fwhm (full width at a half-maximum) of 0.4 eV.

Comment

Silver hydrosols oxidize at air over a period of one day. Further procedures should be performed immediately after preparation of the hydrosol.

B. Transfer of the Silver Nanoparticle Sol into Organic Solvents

A 100 mL proportion of the aqueous silver nanoparticle sol ($2 \cdot 10^{-4}$ M) was transferred into a 2L glass separatory funnel at room temperature in air, and 15 ml of $5.1 \cdot 10^{-3}$ M TOPO (30 mg) in toluene was added. The funnel was closed with a stopper. The addition of TOPO serves two purposes: it brings the silver sol into contact with the immiscible solvent phase by emulsification and also engulfs the particles allowing them to transfer. The mixture was emulsified by strong mixing (shaking funnel by hands) for 25–30 min. During this period, the silver nanoparticles inside the water droplets started to transfer spontaneously to the organic phase. After 30 min of mixing, 22 mg of magnesium sulfate heptahydrate was added to obtain a quantitative transfer of silver particles into the organic phase.[10] The emulsion was mixed for additional 10–15 min to achieve complete dilution of the magnesium salt and separation of the emulsion in two liquid phases. Fast separation in two liquid phases is a necessary condition for self-assembly of silver particles into a film. At the end of this process, the organic phase exhibits a dark yellow color and contains silver nanoparticles, while the residual aqueous phase is clear and colorless.

Characterization

UV/Vis spectra of the organic silver particle sol exhibits a characteristic plasmon absorption with a maximum around 410 nm and fwhm >0.4 eV (Fig. 1). The intensity of plasmon absorption is higher as compared to the aqueous sol because of the higher concentration of the silver particles in toluene.

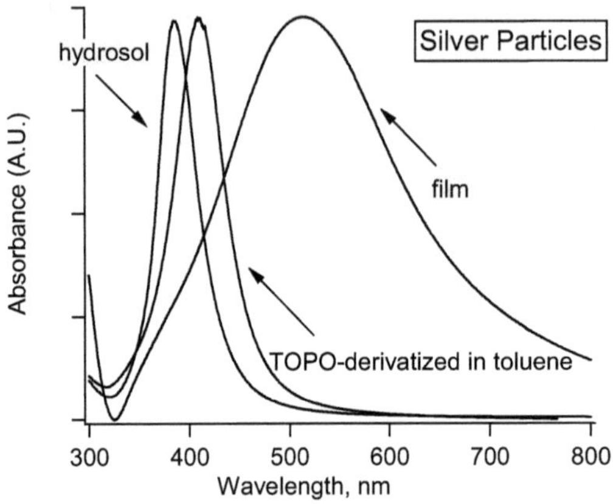

Fig. 1. Normalized optical absorption spectra of silver particles in aqueous solution, TOPO-capped Ag particles in toluene, and of the fine-grain multilayered film.

Comment

The concentration of silver particles in toluene can be lowered by increasing the volume of the added TOPO solution in toluene. However, a lower concentration does not result in self-assembly.

C. Multilayered Film Synthesis

The self-assembly of silver particles to a multilayer film starts during the process of phase separation. Particularly, when toluene contains silver particles (as observed by the dark yellow color of toluene phase), the whole mixture was shaken once again and quickly transferred from the separatory funnel into a 100 mL graduated cylinder in which a microscopy glass slide was immersed in an upright position. The position of the glass slide matches the level of organic phase in the graduate cylinder.

The self-assembly on the glass slide, as well as that on the wall of the cylinder, starts immediately. About 30 sec is usually enough for the deposition of a typical multilayered Ag film. The obtained mirror-like film appears purple when viewed through a glass (transparency mode).

For efficient binding of derivatized silver nanoparticles on the glass slides, their surfaces were thoroughly cleaned. First, the slides were dipped in a 1:1 mixture of acetone and ethanol in an ultra-sound bath for 30 min. Second, the slides were dipped in detergent for the 30 min and then rinsed with water. Third, they were dipped in a 25% solution of sulfuric acid for 30 min and then thoroughly washed with water. The glass slides were finally dipped in *tert*-butyl alcohol and then dried with N_2.

Characterization

The maximum of characteristic plasmon absorption is shifted some 100 nm towards higher wavelengths. Scanning electron microscopy images show that the film consists of 10 nm close-packed particles. Redispersing of particles from the film into toluene results in a silver particle sol, the spectrum of which is virtually identical to that of the original solution used for the formation of multilayered film.

Comment

The thickness of the film can be controlled by the glass immersion time.

References

[1] M. Brust, C. Kiely, J. Colloids and Surfaces A 2002, **202**, 175.

[2] R. P. Andres, T. Bein, M. Dorogi, S. Feng, J. I. Henderson, C. B. Kubiak, W. Mahoney, R. G. Osifchin, R. Reifenberger, Science 1996, **272**, 1323.

[3] S. Chen, R. S. Ingram, M. J. Hostetler, J. J. Pietron, R. W. Murray, T. G. Schaaff, J. T. Khoury, M. M. Alvarez, R. L. Whetten, Science 1998, **280**, 2098.

[4] M. Burst, D. Bethell, D. J. Schiffrin, C. J. Kiely, Adv. Mater. 1995, **7**, 795.

[5] W. P. Wuelfing, R. W. Murray, J. Phys. Chem. B 2002, **106**, 3139.

[6] C. P. Collier, R. J. Saykally, J. J. Shiang, S. E. Henrichs; J. R. Heath, Science 1997, **277**, 1978.

[7] J. J. Shiang, J. R. Heath, C. P. Collier, R. J. Saykally, J. Phys. Chem. B 1998, **102**, 3425.

[8] M. Green, N. Allsop, G. Wakefield, P. J. Dobson, J. L. Hutchinson, J. Mater. Chem. 2002, **12**, 2671.

[9] V. V. Vukovic, J. M. Nedeljkovic, Langmuir 1993, **9**, 980.

[10] H. Hirai, H. Aizawa, J. Colloid Interface Sci. 1993, **161**, 471.

Colloidal Dispersion of Gold Nanoparticles

S. Gross

Abstract Stable sols of gold nanoparticles with an average diameter of about 13 nm were prepared from hydrogen tetrachloroaurate (prepared from bulk metallic gold) and trisodium citrate dihydrate. The organic salt acts as reducing as well as stabilizing agent for the gold nanoparticles, having an average diameter of 13±4 nm. The deep purple colloidal suspension is stable for several weeks. UV-Vis absorption spectra shows the typical surface plasmon resonance band of nanosized gold with an absorption maximum at 521.5 ± 0.5 nm.

Classification

form:	colloid
function:	pigment
preparation:	reduction of aqueous metal solution
composition:	citrate-stabilized Au nanoparticles

Introduction

Gold colloids have been known since ancient times for their fascinating properties and colors, which nowadays can be related to the presence of metal nanoparticles.[1] Starting from the pioneering investigations of M. Faraday[2] and W. Ostwald,[3] gold nanoparticles have been the topic of much interest due to their easy preparation and high stability.[4] An extensive review on their synthesis, properties and applications has been recently published.[1] Nucleation and growth of colloidal gold was thoroughly investigated by Turkevich *et al.*[5,6,7]

Gold nanoparticles are well-known for producing a strong optical response (plasmon) due to the excitation of free electrons at the metal surface which results in typical surface plasmon resonance (SPR) bands.[8] These outstanding optical properties of gold nanoparticles make them suitable for several uses. Furthermore, gold nanoparticles are appealing systems for their invaluable chemical, supramolecular, recognition and catalytic properties.[1] Au nanoparticles-oligonucleotide conjugates are currently attracting great interest because of their potential use in DNA detection. Gold colloids have also found application as non-cytotoxic labels, biolabels and in optical biosensors.[9] Colloidal gold, labeled to various biological materials like lectins, antibodies, antigens, enzymes, or lipoproteins allows to observe these systems by transmission or scanning electron microscopic methods.[9]

Two main approaches can be used to generate metal colloids, the first based on dispersion of larger particles (*dispersion method*), the second relying on the condensation of smaller units (*reduction method*).[9] While the former affords only unstable sols, characterized by particles with a wide size distribution, the latter allows to prepare stable sols through reduction of metal salts in solution.[9] Several routes have been proposed to prepare stable suspensions of gold nanoparticles,[1,6,10,11,12,13] mainly based on reduction of Au(III) derivatives.

For example, gold colloids are easily prepared by reduction of $HAuCl_4$ in diluted aqueous solution with citric acid or trisodium citrate. A very narrow size distribution can be obtained when the latter reducing agent is used which is oxidized to carbon dioxide in the course of the reaction. Carbon dioxide is formed after different oxidative steps; intermediates and by-products such as acetone dicarboxylic acid have been reported.[5]

The method described here allows obtaining very stable gold nanoparticles with a diameter of about 13 nm and a quite narrow size distribution, starting from a solution of hydrogen tetrachloroaurate and trisodium citrate. Trisodium citrate acts both as reducing as well as stabilizing agent.

$$6\ Au^{3+} + C_6H_5O_7^{3-} + 15\ OH^- \longrightarrow 6\ Au + 6\ CO_2 + 10\ H_2O$$

The proposed method presents, with respect to other routes, several advantages, mainly related to i) easy synthetic procedure, ii) reproducibility of the method iii) stability of the prepared sol. The aqueous medium used is another advantage, since water solvates both reagents very well.

Variation of the experimental conditions allows, in principle, tailoring the particles size up to 900 nm.[9] In particular, Turkevich *et al.* [5] have extensively described the effect of various parameters, such as the temperature, amount of citrate added or the dilution of the solution, on the formation of colloidal gold. Lowering the temperature at which the sodium citrate solution is added to the hydrogen tetrachloroaurate solution by 10°C, increases the time required for completion of the reaction by a factor of two. The amount of citrate added or the dilution of the solution can dramatically affect the average size and size distribution of the gold nanoparticles. The latter is additionally depending on the relative rates of nucleation and growth.[5]

However, although variation of the above mentioned parameters over a large range yields sols of nanoparticles of different size and size distribution, the general outcome of the procedure (colloidal gold) remains unaffected. It should furthermore be pointed out that by using the described procedure, the variation of the citrate/hydrogen tetrachloroaurate(III)·$3H_2O$ molar ratio within a limited range does not dramatically affect the size and size distribution. However, the prepared gold colloids are only stable in solution, because they are protected by ligand molecules and electric charges preventing coagulation.

Materials

- Metallic gold foil purchased from Nuova Franco Suisse Italia
- Hydrogen chloride, HCl, 37% (Aldrich), used as received
- Nitric acid, HNO_3, 65% (Carlo Erba), used as received
- Sulfuric acid, H_2SO_4, 98% (Carlo Erba), used as received
- Sodium citrate dihydrate, $HOC(COONa)(CH_2COONa)_2{\cdot}2H_2O$, 99.0% ACS reagent (Merck) used as received

Safety and Disposals

Safety and handling instructions for the chemicals, especially those involved in the preparation of $HAuCl_4$ (strong acids), are reported in the corresponding materials safety data sheets (MSDS). The employed chemicals should be handled with care and with protective gloves. Preparation of $HAuCl_4$ should be carried out in well ventilated areas and under an aspirated fume hood due to the development of toxic nitric vapors. Gloves and safety glasses should be worn when working with the precursor solutions as well as with the colloidal sol.

Procedures

All the procedures were performed in air, at room temperature and atmospheric pressure, using de-ionised water.

A. Preparation of $HAuCl_4{\cdot}3H_2O$

Synonyms: Hydrogen tetrachloroaurate trihydrate, tetrachloroauric acid trihydrate; chloroauric acid trihydrate.

The reported procedure allows preparing about 1.3 g of $HAuCl_4{\cdot}3H_2O$ and requires about one week for the preparation and about two weeks for the crystallization of the product.

For all concentration/evaporation steps, a wide flask (Petri dish) should be used, because otherwise condensation of the acid will require considerably more time.

(1) About 0.7 g of metallic gold is cut in very small pieces (~ 1 × 2 mm) and put into a 250 ml beaker. 70 ml of aqua regia (75% v/v HCl, 25% v/v HNO_3) are slowly added.
(2) The mixture is stirred and gradually heated to 50°C. When dissolution of gold slows down, the temperature is gradually increased to 70-80°C.
(3) Once metallic gold is completely dissolved (after about 2 h), the solution is continuously heated until it is concentrated to 30 ml.
(4) HCl is slowly added to the hot solution, until brown nitric vapors are completely eliminated and the volume is about 60 ml.
(5) The procedure described in 3. and 4. is repeated about 5 times, until, after addition of HCl, no brown nitric vapor is developed.
(6) The solution is concentrated (by heating) to 30 ml.
(7) Bidistilled de-ionised water is added to the solution under stirring which is kept at about 70-80°C, until a volume of about 50 ml is obtained and until acid vapors are completely absent. The presence of acid vapors is checked by a litmus paper (pH indicator) which is put into the vapors; this operation (addition of water, concentration of the solution under heating) is repeated until pH 7 is reached.
(8) The magnetic stirrer is removed from the beaker and the solution is concentrated to 15 ml by heating at 70°C;
(9) The solution is cooled to room temperature.
(10) The beaker is put into a desiccator; on the bottom of the desiccator a crystallizing dish containing concentrated H_2SO_4 is placed;
(11) The desiccator is put in vacuum by using a water pump and protected from sunlight by an aluminum foil;
(12) The vacuum is periodically checked and the desiccator is left standing for about 15 d, until crystals of an intense yellow color are formed.

B. Preparation of Gold Nanoparticles Colloidal Suspension

The reported procedure allows the preparation of a stable suspension of monodisperse gold nanoparticles having a diameter of about 13 nm.

A 1 mM aqueous solution of $HAuCl_4$ is prepared by dissolving 0.39 g of $HAuCl_4 \cdot 3H_2O$ in 1 L of de-ionised water, and a 38 mM solution of sodium citrate dihydrate by dissolving 11.41 g in 1 L de-ionised water. An amount of 20 ml of the $HAuCl_4$ solution is put in a 50 ml beaker equipped with a magnetic bar. The solution is heated to the boiling point and then 2 ml of the citrate solution are added (Au : citrate molar ratio 1:3.8) under stirring. The yellow color of the solution originating from hydrogen tetrachloroaurate first turns colorless. The solution remains clear for about 10 sec and then turns grayish-blue. After about 1 min a deep wine-red sol is obtained. No further change of color upon prolonged boiling is observed. Deionised water is added dropwise to keep the volume at 22 ml. The solution is then cooled to room temperature. The pH of the final solution is about 6.7.

Characterization

A purple red, stable sol is obtained.

UV-Vis characterization: The UV-Vis absorption spectra of the colloidal sol were acquired in the 200-800 nm wavelength range, using a 2 nm spectral band width and a 0.5 nm data interval. The absorption maximum is at 521.5±0.5 nm, in agreement with values reported in literature (Fig. 1).[1,9,12]

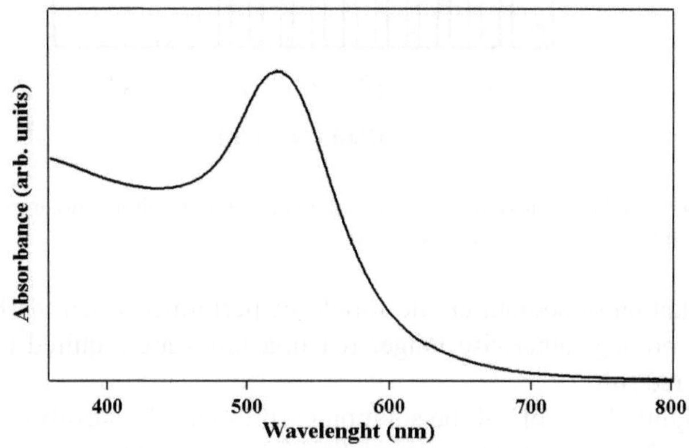

Fig. 1. UV-Vis spectrum of the gold colloid (molar Au: citrate ratio1: 3.8).

Dynamic light scattering measurements (Particle Sizing Systems Nicomp Model 370 correlator equipped with a thermostated cell holder and a Spectra Physics Series 2016 Ar laser operating at 488 nm): Hydrodynamic particles diameters were obtained from cumulated fits of the autocorrelation functions at 90° scattering angle. DLS measurements were performed on the gold sol twice diluted in Milli-Q water. The sol was filtered through a Sartorius Minisart single use filter (0.2 μm) before measurements. The average particle size (number-weighted) was 12 ± 4 nm (Fig. 2).

Comments

(1) The prepared sol is stable for several weeks.
(2) The replacement of sodium citrate by other reducing agents would result in other nanoparticles sizes and dispersions.

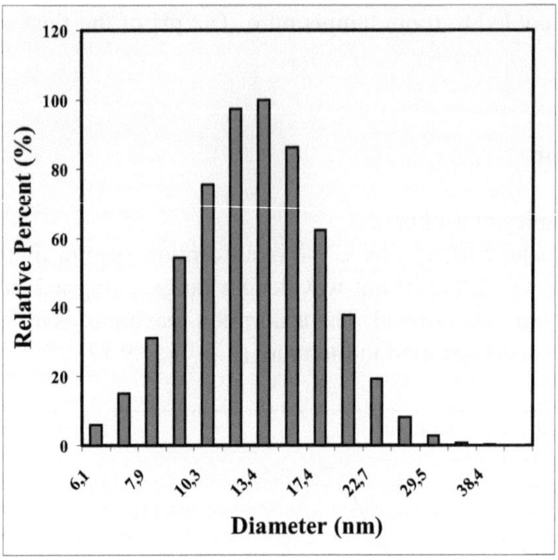

Fig. 2. Number-weighted Gaussian analysis of the gold sol (Au: citrate molar ratio1: 3.8) as determined by light scattering measurements.

(3) The addition of sodium citrate should be performed when the $HAuCl_4$ solution is boiling, otherwise longer reaction times are required to achieve the purple red sol.

(4) The slightly basic pH of the solution ensures that the adsorbed citrate groups on the gold nanoparticles are completely deprotonated.

(5) The features of the colloidal sols obtained by this procedure are affected by experimental parameters, such as the temperature: lowering the temperature at which trisodium citrate is added to the chloroauric acid solution by 10°C increases the time required to observe the deepening of the solution color, which indicates the completion of the reaction, by a factor of 2.[1]

(6) The preparation of colloidal gold by this procedure starting from hydrogen tetrachloroaurate and trisodium citrate solutions is easy, safe, not dangerous, and can be performed in a reproducible way also by undergraduate chemistry students.

(7) Addition of strong electrolytes to the gold colloids results in their fast coagulation and precipitation.

References

[1] M. C. Daniel, D. Astruc, Chem. Rev. 2004, **104**, 293, and references therein.
[2] M. Faraday, Philos. Trans. 1857, **147**, 145.
[3] W. Ostwald, Kolloid Z. 1909, **4**, 5.

[4] M. A. Hayat, Collodial Gold, Principles, Methods and Applications, Vol. 1; Academic Press, New York, 1989. G. Schmid, Clusters and Colloids. From Theory to Application, VCH, Weinheim, 1994.
[5] J. Turkevich, P. C. Stevenson, J. Hillier, Discuss. Faraday Soc. 1958, **11**, 55.
[6] B. V. Enüstün, J. Turkevich, J. Am. Chem. Soc. 1963, **85**, 3317.
[7] D. Beischer, F. Krause, Angew. Chem. 1938, **51**, 331.
[8] U. Kreibig, M. Vollmer, Optical Properties of Metal Clusters, Springer, Berlin, 1995.
[9] G. Schmid, Chem. Rev. 1992, **92**, 1709.
[10] J. Turkevich, J. Hillier, Anal. Chem. 1949, **21**, 475.
[11] G. Schmid, B. Corain, Eur. J. Inorg. Chem. 2003, 3081.
[12] A. D. McFarland, C. L. Haynes, C. A. Mirkin, R. P. Van Duyne, H. A. Godwin, J. Chem. Educ. 2004, **81**, 544A.
[13] S. L. Cumberland, G. F. Strouse, Langmuir 2002, **18**, 269.

One-dimensional Nanorods and Nanowires

Hao Ming Chen and Ru-Shi Liu

Abstract Gold nanorods and nanowires were fabricated by controlling the volume of growth solution. Shape evolutions ranging from rice-like nanoparticles to 1-D rods were observed. The addition of growth solution can control the length of nanorods to 2 μm, and nanorods with aspect ratios of up to ~70 can be obtained.

Classification

form:	colloid
function:	pigment
preparation:	reduction of aqueous metal solution
composition:	cetyltrimethylammonium bromide-capped Au nanorods and nanowires

Introduction

Numerous characteristics of nanomaterials depend on size and shape, including their catalytic, optical, and physical properties.[1-3] A number of chemical approaches have been actively explored to process metal into 1-D nanostructures.[4] Gold nanorods have been synthesized by electrochemical reduction method in presence of cetyltrimethylammonium bromide (CTAB) [2,5] and by seed-mediated growth in a surfactant template.[3] The growth mechanism of 1-D gold nanoparticles in the presence of CTAB has been extensively investigated in the literature (e.g., growth direction, micelles properties, optical absorption spectra, effect of pH).[6-10] It is vitally important to study the fundamentals of gold nanorods and nanowires growth because the understanding of this aspect is a guide to new materials design and more sophisticated synthetic methods. Here, by successive addition of growth solution to seed solution, the shape and length of product was controlled.

Materials

- Hydrogen tetrachloraurate(III) hydrate, $HAuCl_4$ (Across), used as received
- Trisodium citrate dihydrate, $HOC(COONa)(CH_2COONa)_2$, 99% (Across), used as received
- Silver nitrate, $AgNO_3$, 99% (Across), used as received
- Cetyltrimethylammonium bromide, $C_{19}H_{42}NBr$, 99.0% (Across), used as received
- Ascorbic acid, $C_6H_8O_6$, 99% (Across), used as received

Safety and Disposal

Safety and handling instructions for the chemicals, especially those involved in the preparation of $HAuCl_4$ (strong acids), are reported in the corresponding materials safety data sheets (MSDS). The employed chemicals should be handled with care and with protective gloves.

Procedures

All the procedures were performed in air, at room temperature and atmospheric pressure, using de-ionized water.

Preparation of Gold Seeds

An aqueous 1% trisodium citrate solution (0.35 mL) was added into 10 mL of an aqueous 0.25 mM $HAuCl_4$ solution. After the solution was stirred for 3 min, 0.3 mL of an ice-cold, freshly prepared aqueous 0.01 M $NaBH_4$ solution was added, followed by stirring for 5 min. The seed solution was kept at room temperature for ~ 2 h and was used further.

Preparation of Growth Solution

0.08 M CTAB and 250 μM $HAuCl_4$ aqueous solution was prepared as growth solution. The solution was heated to 40°C while stirring to dissolve the CTAB. The solution was then stored at 27°C until cooling to room temperature.

Preparation of Gold Nanorods and Wires

Amounts of 0.01, 0.1, 1 and 10 mL of freshly prepared 10 mM ascorbic acid (AA) solutions were mixed with 0.2, 2, 20 and 200 mL of growth solutions, respectively. Next, 0.03 and 0.3 mL $AgNO_3$ (0.025 mM) solution were added to 0.2 and 2 mL of growth solution, respectively. The compositions of these four solutions are shown below. These four colorless solutions were added to the 0.02 mL of gold seed solution one by one at 40 sec intervals.

Solution	Composition
I	0.2 mL growth solution, 0.01 mL of AA solution, 0.03 mL $AgNO_3$
II	2 mL growth solution, 0.1 mL of AA solution, 0.3 mL $AgNO_3$
III	20 mL growth solution, 1 mL of AA solution
IV	200 mL growth solution, 10 mL of AA solution
V	2000 mL growth solution, 100 mL of AA solution

Characterization

TEM characterization: Fig. 1 shows a typical transmission electron microscopy (TEM) image of the gold nanoparticles and rods. The inset shows a SEM image of rice-like nanoparticles (Fig. 1A and B) as solutions I (Fig. 1A) and (I+II) (Fig. 1B) were added. When growth solutions (I + II + III) (Fig. 1C) and (I + II + III + IV) (Fig. 1D) were added, nanorods were observed with average lengths being ~550 nm and 1.4 μm, respectively. Nanorods with aspect ratios of up to ~40 can be obtained. The TEM and SEM analysis clearly indicates that the shape evolves from rice-like nanoparticles to 1-D rods. Note that the nanorods could be expanded up to 2μm and nanowires with the aspect ratios of up to 70 (Fig. 2, the inset shows electron diffraction pattern of gold nanowires).

UV-Vis absorption characterization: The absorption spectra are shown in Fig. 3. It is well known that the surface plasmon absorption spectra of gold nanorods are characterized by two bands, the shorter wavelength band is attributed to the transverse surface plasmon resonance and another absorption band appears at longer wavelength, which corresponds to the longitudinal surface plasmon resonance. The rice-like nanoparticles exhibit transverse and longitudinal plasmon bands in the visible region of the spectrum.

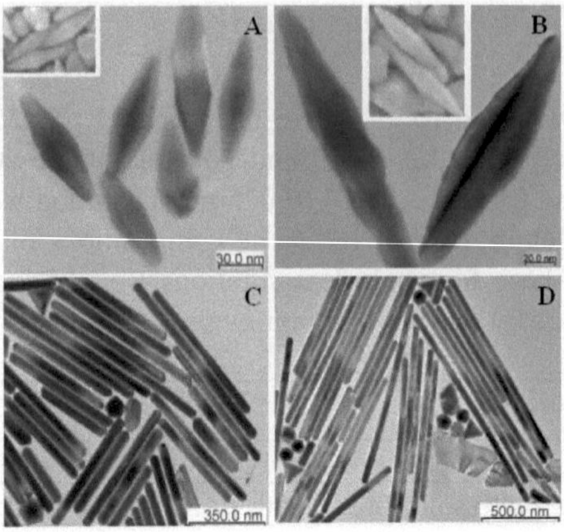

Fig. 1. TEM images of gold nanoparticles synthesized by this method. A, B, C, and D represent particles after addition of seed solutions (I), (I+II), (I+II+III) and (I+II+III+IV), respectively. The inset shows the SEM image of the corresponding samples.

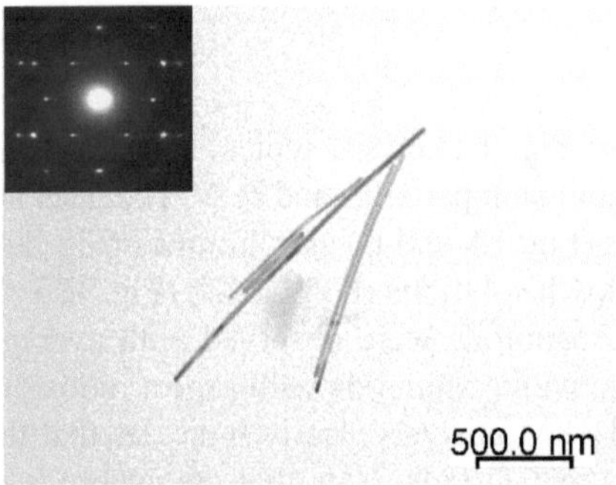

Fig 2. TEM images of nanorods after solution (I+II+III+IV+V) was introduced into the seed solution. The inset shows electron diffraction pattern of gold nanowires.

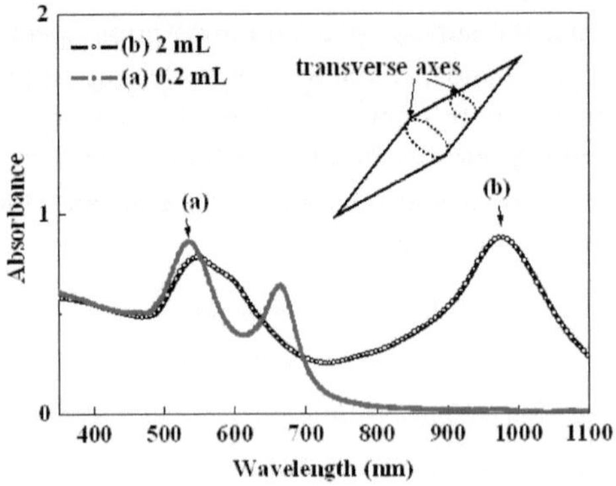

Fig. 3. Extinction spectra of Au rice-like nanorods (a) after (I) solution was added, (b) after (I+II) growth solution was added.

Comments

(1) The prepared solution of the 1-D nanorods/wires is stable for several weeks.
(2) The replacement of CTAB by other reducing agents would result in other nanoparticles shapes.
(3) The addition of ascorbic acid should be quick because the gold atom would aggregate and grow if ascorbic acid is introduced slowly.
(4) The preparation of colloidal gold by this procedure starting from hydrogen tetrachloroaurate and trisodium citrate solutions is easy and safe.
(5) Addition of strong electrolytes to the gold colloids results in their fast coagulation and precipitation.

References

[1] C. J. Murphy, Science 2002, **298**, 2139.
[2] Y.-Y. Yu, S.-S. Chang, C.-L. Lee, C. R. Chris. Wang, J. Phys. Chem. B 1997, **101**, 6661.
[3] N. R. Jana, L. Gearheart, C. J. Murphy, Adv. Mater. 2001, **13**, 1389.
[4] Y. Sun, B. Gates, B. Mayers, Y. Xia, Nano Lett. 2002, **2**, 165.
[5] S.-S. Chang, C.-W. Shih, C.-D. Chen, W.-C. Lai, C. R. Chris. Wang, J. Phys. Chem. B 1999, **15**, 701.
[6] C. J. Johnson, E. Dujardin, S. A. Davis, C. J. Murphy, S. Mann, J. Mater. Chem. 2002, **12**, 1765.
[7] M. Tornblom, U. Henriksson, J. Phys. Chem. B 1997, **101**, 6028.
[8] N. R. Jana, L. Gearheart, S. O. Obare, C. J. Murphy, Langmuir 2002, **18**, 922.
[9] T. K. Sau, C. J. Murphy, J. Am. Chem. Soc. 2004, **126**, 8648.
[10] B. D. Busbee, S. O. Obare, C. J. Murphy, Adv. Mater. 2003, **15**, 414.

Monolithic Tin-doped Silica Glass

N. Chiodini, F. Morazzoni and R. Scotti

Abstract Monolithic and transparent Sn-doped SiO_2 glasses, where Sn atoms replaced Si centers in the SiO_2 network, were prepared by a new sol-gel route by using tetraethoxysilane (TEOS) and dibutyltindiacetate (DBTDA) as precursors. The maximum amount of Sn doping was 1.40 wt % $SnO_2/(SnO_2+SiO_2)$ (corresponding to 0.55 mol %). At higher tin content ($\geq$ 1.60 wt %, corresponding to 0.64 mol %) crystalline nanosized particles of SnO_2 (6-10 nm) segregated in silica matrix.

Classification

form:	monolithic glass
function:	optoelectronics
preparation:	sol-gel processing
composition :	Sn-doped SiO_2

Introduction

Sn-doped SiO_2 glass is an interesting material for technological applications in optoelectronics, e.g. Bragg gratings in optical fibers or wave guides,[1] owing to its high UV photosensitivity.[2, 3] The photosensitivity is the property of changing the refractive index of the material by exposure to an optical radiation. Bragg gratings are permanent refractive index gratings, produced by interference of visible or UV waves within Ge-doped silica fibers. Different models have been proposed to explain the origin of the photorefractivity, e.g. the color-centre model [4] and the densification model,[5] but the process is generally associated with the presence of defect centers related to Ge atoms located in tethrahedral sites of silica network.[6] Many attempts were made to dope silica with elements other than Ge (e.g. Al, P, rare earths) in order to enhance the sensitivity of gratings [3] but a significant improvement was obtained with Sn doping. As a matter of fact, it was demonstrated that Sn-doped SiO_2 fibers showed a photosensitivity comparable with the more common Ge-doped SiO_2 fibers but containing a Sn amount nearly two orders of magnitude lower than Ge.[1,7]

Thus it is important to develop a synthesis procedure of monolithic Sn-doped silica glass with the higher concentration of Sn atoms substitutionally introduced in SiO_2 network.

The incorporation of Sn in SiO_2 is difficult due to the possible crystallization of SnO_2, which occurs at very low Sn concentration and to the volatility of SnO_2, a drawback for the high temperature preparation methods of glasses. Sn-doped SiO_2 fibers with $\approx$0.15 mol % were produced via modified chemical-vapor deposition (MCVD) [1] but higher Sn concentrations were obtained only in the presence of codopant or glass modifiers.[8, 9]

The sol-gel method, via hydrolysis and condensation of molecular precursors at low temperatures, was used in this procedure to prepare doped SiO_2 glasses with a substitutional Sn content higher than that obtained by high temperature methods. The sol-gel method allows easier control of the composition in the sol-phase and the thermal treatments of glass densification process. Furthermore, preforms for fibers and films for planar waveguides can be produced. However, when the simultaneous hydrolysis of different metal precursors occurs, different reaction rates could lead to a lack of homogeneity in the gels. For this reason dibutyltindiacetate (DBTDA) was chosen as the tin precursor instead of tin alkoxides, which are highly reactive with water and easily form hydroxo- or oxohydroxo precipitates. DBTDA is used for curing silicones as it can give crosslinking reactions between the silanol groups of low molecular weight silicone.[10]

The synthesis procedure to prepare transparent, monolithic Sn-doped SiO_2 glasses, where Sn atoms replaced Si centers in the SiO_2 network, is described. The limit in doping SiO_2 with Sn is 1.40 wt % $SnO_2/(SnO_2+SiO_2)$, corresponding to 0.55 mol %. At higher tin content ($\geq$ 1.60 wt %, corresponding to 0.64 mol %) the segregation of crystalline nanosized particles of SnO_2 in silica matrix occur.

The substitutional position of Sn in SiO_2 tetrahedral sites was demonstrated by the presence of paramagnetic E'-Sn centers, a three coordinated tin center with an unpaired spin in a sp^3-like orbital, formed by X-ray irradiation and detected by Electron Paramagnetic Resonance (EPR).[11]

The properties of the material,[12,13] the study of the hydrolysis and condensation reactions [14] and the study of the thermal evolution during sintering process [15] are reported elsewhere.

Materials

- Dibutyltindiacetate (DBTDA), Fluka > 98.0 %.
- Tetraethoxysilane (TEOS), Stream Chemicals 99.9999%.
- Ethanol, HPLC grade reagent; water Mill-Q

Safety and Disposal

Safety and handling instructions for the chemicals are found in the corresponding materials safety data sheets (MSDS). DBTDA and TEOS are toxic by inhalation, in contact with skin and if swallowed. Both compounds are combustible; TEOS is moisture sensitive.

Procedures

Tin doped silica xerogels and glasses were prepared in the range of compositions 0.01–10.0 wt % $SnO_2/(SiO_2+SnO_2)$ corresponding to 0.004–4.24 mol %.

Sol Preparation

The sol-phases were prepared in flasks or directly in polypropylene containers used for the gelation and drying steps (see next paragraph), adding successively, under stirring, 7.00 ml (120 mmol) of ethanol, 2.50 ml (11.2 mmol) of TEOS, the appropriate amount of DBTDA and, finally, 1.50 ml (83.3 mmol) of water. The molar ratio Si:H_2O:EtOH was 1:7.4:10.7. Attention must be paid to TEOS transfer into the flask as it is moisture sensitive. The amount of added DBTDA depended on the desired tin content. In more concentrated samples, pure DBTDA was added. Example: for the 10.0 wt% (4.24 mol%) sample, add 0.134 ml of DBTDA (0.496 mmol). In more diluted samples, an ethanol solution of DBTDA was added. Examples: for the 1.0 wt% (0.401 mol%) sample, add 0.109 ml of ethanol:DBTDA solution (100:1 v:v) corresponding to 0.045 mmol of DBTDA. The amount of the ethanol added with the DBTDA solution did not significantly change the Si:H_2O:EtOH ratio.

Sol-Gel Transition and Drying

The containers (diameter 6 cm, height 6 cm) were sealed by a polyethylene film and put into a thermostatic chamber at 40±1 °C. Three small holes in the film (about 1 mm diameter) allowed for solvent evaporation.

Gelation times depended on the tin content and ranged from 72 h (0.01 wt%) to 24 h (10.0%). Gelation times were taken when the sol-phase lost its liquid characteristics and transformed into a continuous phase holding the shape of the container. The absolute values had an uncertainty of about ± 30% but the relative trend vs. tin content was confirmed by repeated experiments.

After gelation the samples were slowly dried in a thermostatic chamber at 40±1°C for 15 d.

Xerogels can already be obtained in 3-5 d but a longer aging (10-15 d) favors the reproducibility of the xerogel properties. The dried gels (xerogels) were monolithic, transparent and colorless in the whole range of compositions.

Thermal Treatment and Glass Formation

Glasses were obtained from the xerogels by sintering process A, performed in an electrical furnace equipped with a temperature controller and a tubular quartz camera connected with the vacuum system (rotary pump and Pirani gauge to measure pressure) and the gas stream source.

Sintering Process A

a) The temperature was increased (10 K h^{-1}) from room temperature to 378 K under a stream of oxygen (50 ml min^{-1}), and the samples were then held at this temperature for 4 h;
b) the temperature was increased (10 K h^{-1}) from 378 K to 723 K under a stream of oxygen (50 ml min^{-1}), and the samples were maintained at 723 K for 48 h;
c) the temperature was increased (4 K h^{-1}) from 723 K to 1323 K under a stream of oxygen (50 ml min^{-1});
d) the temperature was decreased (70 K h^{-1}) from 1323 K to room temperature under a stream of oxygen (50 ml min^{-1}).

Monolithic plates of about 15 mm diameter and 2 mm thick were obtained. The slow heating rates of the sintering process were necessary to prevent glass cracking.

The glasses were colorless up to 1.40 wt % $SnO_2/(SnO_2+SiO_2)$ (corresponding to 0.55 mol%). At higher amounts of the dopant (≥ 1.60 wt %, corresponding to 0.64 mol%) they became yellow, the color intensity increasing with the tin content. In the transparent glasses Sn atoms replaced Si centers in the SiO_2 network; in yellow glasses, at higher tin content, particles of SnO_2 segregated in silica (see Characterization).

The samples were held at 378 K to account for the loss of physisorbed water from xerogel, and at 723 K to remove most of the chemisorbed hydroxy groups and to combust the organic entities completely.

An oxygen stream must be used at least up to 723 K to completely burn and eliminate organic groups. At higher sintering temperatures, the oxygen partial pressure influences the glass defectivity. Treatment in pure oxygen prevented the formation of oxygen-deficient defects, detected by their characteristic absorption at about 4.9 eV.[15]

Sintering Process B

This sintering process is an example for the treatment at low partial pressure of oxygen. It was performed in vacuum (1.33 Pa) from 723 K to 1123 K and under a stream of He:O_2 (1.0 vol %) at T>1123 (in vacuum at T>1123 glass darkening occurs due to tin reduction).

a) Same as the sintering process A;
b) same as the sintering process A;
c) the temperature was increased (4 K h^{-1}) from 723 K to 1123 K in vacuum (1.33 Pa);
d) the temperature was increased (4 K h^{-1}) from 1123 K to 1323 K under a stream of He:O_2 (1.0 vol %, 50 ml min^{-1}) and then the temperature was maintained at 1323 K for 20 h;
e) the temperature was decreased (70 K h^{-1}) from 1323 K to room temperature under a stream of He:O_2 (1.0 vol %, 50 ml min^{-1}).

Monolithic glasses were obtained with the same characteristics as glasses produced by sintering method A except the presence of oxygen defects (see UV-Vis spectroscopy in Characterization section).

Characterization

BET surface area measurements: The measurements were performed on a Coulter SA 3100 instrument after outgassing the samples at 373 K for 60 min. The surface area decreased markedly with increasing treatment temperature. The 0.01 %, xerogel may serve as an example: before thermal treatment, 687 m^2g^{-1}; sintered at 673 K, 571 m^2g^{-1}; at 1023 K the surface area was below the detection limit (< 1 m^2g^{-1}).

AES-ICP analysis: Inductively Coupled Plasma Atomic Emission Spectroscopy (AES-ICP) analysis of tin was performed with a Jobin-Yvon 38 instrument. About 60 mg of the glass were first dissolved in a solution 48% w/w of HF (about 10 ml). After addition of conc. H_2SO_4 (0.5 ml), the solution was heated to eliminate fluorides, then diluted with H_2O Mill-Q in a volumetric flask (10 ml) and the tin content measured. Tin analyses performed on densified glass samples revealed the same Sn content as the sol precursors showing that no loss of tin occurred during the thermal treatment.

UV-Vis spectra were recorded on a Cary 2300 Varian spectrophotometer. The optical absorption spectra of the yellow glass ($\geq$ 1.60 wt %) showed an absorption edge at about 3.6 eV, which correspond to band-to-band transition of SnO_2.[16] The colorless glasses ($\leq$1.40 wt%) showed a tail of a strong absorption at higher energy (about 6 eV) typical of doped silica [17] and a band at 4.9 eV associated with doping-induced oxygen defects (Fig. 1).[18] This band was not observed for glasses which underwent the sintering process A.

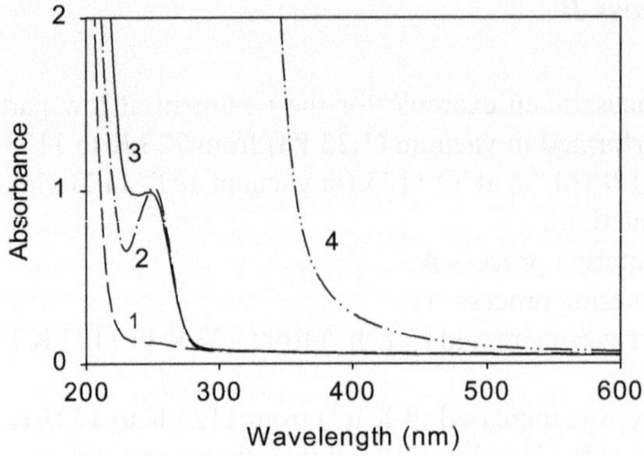

Fig. 1. UV absorption spectra of Sn-doped silica glasses with different tin content: (1) 0.5 wt% sintered by process A (see text); (2) 0.5 wt% sintered by process B; (3) 1.0 wt% and (4) 10.0 wt% both sintered by process B.

X-Ray diffraction: The powder X-ray diffraction (XRD) patterns were obtained at ambient conditions with a Siemens D 500 diffractometer using monochromatic Cu Kα radiation (λ = 1.5418 Å). The average crystallite size D of SnO_2 particles dispersed in silica glass was calculated by the reflection from the (110) plane according to the Scherrer formula, D =0.9λ/(ß cosθ), where λ is the X-ray wavelength, 2θ is the diffraction angle. $\beta = \sqrt{(\beta_n^2 - \beta_s^2)}$ is the corrected halfwidth in radians, with β_n the observed (110) reflection halfwidth of SnO_2 in Sn-doped silica samples and β_s the halfwidth in a standard sample of SnO_2 powder. The standard sample of SnO_2 was obtained by sintering a powdered xerogel of pure SnO_2 at 1323 K under a stream of oxygen for 24 h.

No crystalline phase was observed at tin contents lower than 1.40 wt%. On the other hand, the diffraction patterns of the yellow glasses showed the cassiterite crystalline structure (Fig. 2).[19] The peak intensities increased with the tin content. The average size of the SnO_2 grains, calculated by the Scherrer formula, was about 6 nm in 3.0 wt% glass and 10 nm in 10.0 wt%.

Raman spectroscopy: Measurements were performed on a Labram Dilor microRaman spectrometer excited by a He-Ne laser in back-scattering configuration. Relative peak positions were determined with a final precision of about 1 cm^{-1}. Relative peak intensities were analysed by normalising the spectra at the intrinsic ω_3 mode of SiO_2 at 800 cm^{-1}, which is well separated from all other peaks.

Raman spectroscopy allowed monitoring the glass densification during the thermal treatment. After the sintering process at 1323 K, the Raman spectra (300 - 1300 cm^{-1}) of Sn-doped silica glasses showed a pattern where the intensity ratio of the main SiO_2 bands was typical of densified pure silica glass (Fig. 3). The main bands are the D_1 and D_2 peaks at 490 and 610 cm^{-1}, attributed to symmetric stretching modes of vibrationally isolated four- and three-membered rings of SiO_2 tetrahedra;[20] the band at 440 cm^{-1}, attributed to symmetric stretching ω_1 mode of SiO_2;[21,22] the bands at 800 cm^{-1} (ω_3 mode), 1060 cm^{-1} (transverse optic (TO)

ω_4 mode) and 1190 cm^{-1} (longitudinal optic (LO) ω_4 mode).[21] Xerogels sintered at temperature below 1323 K showed different intensity ratios between these peaks.[13]

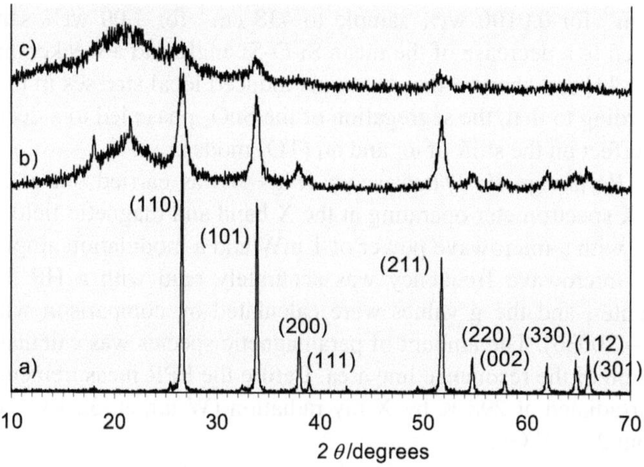

Fig. 2. XRD patterns of a) SnO_2 powder reference (*hkl* Miller indices are indicated); b) Sn-doped silica glass (10.0 wt %); c) Sn-doped silica glass (3.00 wt%).

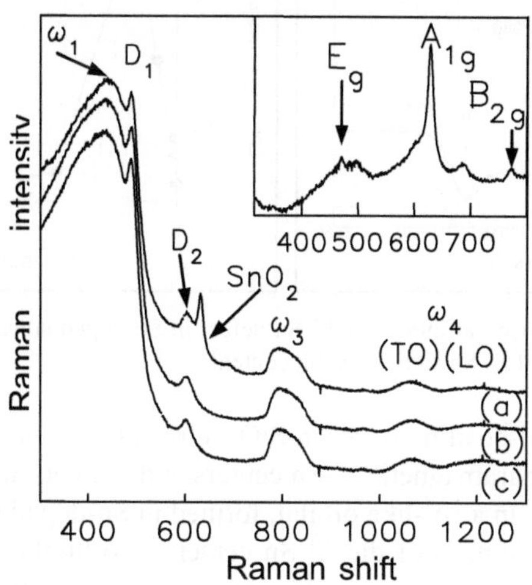

Fig. 3. (a): Raman spectra of Sn-doped silica glasses with tin content (a) 10.0 wt %, (b) 0.500 wt % and (c) pure SiO_2. Inset: difference between spectra (a) and (c).

Features of crystalline SnO_2 were the narrow and intense peak at 630 cm^{-1} (A_{1g} mode) and the less intense peaks at 476 cm^{-1} (E_g mode) and at 782 cm^{-1} (B_{2g} mode),[23] which can be detected only by difference between the spectra of doped and pure silica and pure silica (inset in Figure 3). In glasses with tin content lower that 1.40 wt%, no peaks attributable to vibrational modes directly involving Sn atoms were observed.

Raman also confirmed that tin atoms substituted silicon in the silica network. In glasses with tin content lower than 1.40 wt%, a slight shift of the intrinsic modes ω_1 and ω_4 (TO) with the increasing of dopant amount occurred (e.g. ω_1 shifted from 434 cm^{-1} for 0.0100 wt% sample to 438 cm^{-1} for 1.00 wt% sample). This shift is related to a decrease of the mean Si-O-Si angle and a weakening of the Si-O bonds [13,24] and showed that tin atoms induced local stresses in the silica network. According to that, the segregation of the SnO_2 phase led to a decrease of the Sn-doping effect on the shift of ω_1 and ω_4 (TO) modes.

Electron Paramagnetic Spectroscopy (Fig. 4) was carried out at 298 K by a Bruker EMX spectrometer operating at the X band and magnetic field modulation of 100 kHz, with a microwave power of 1 mW and a modulation amplitude of 0.3 Gauss. The microwave frequency was accurately read with a HP 53131A frequency counter, and the g values were calculated by comparison with a DPPH standard (g=2.0036). The amount of paramagnetic species was calculated by double integration of the resonance line area. Before the EPR measurements, the samples were irradiated at 298 K by X-ray radiation (W target, 32 kV, 20 mA) at a dose of about 2 x 10^4 Gy.

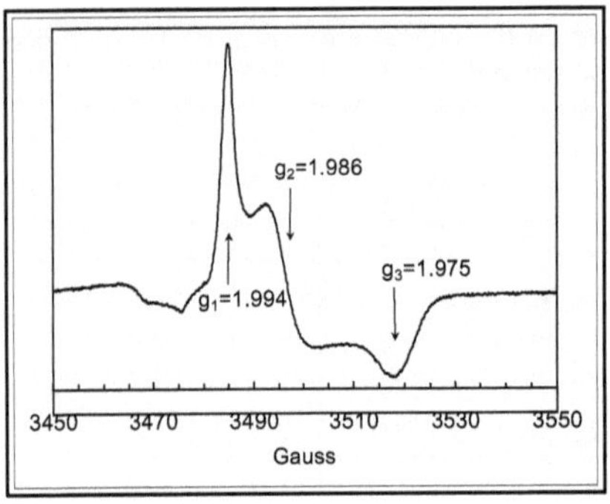

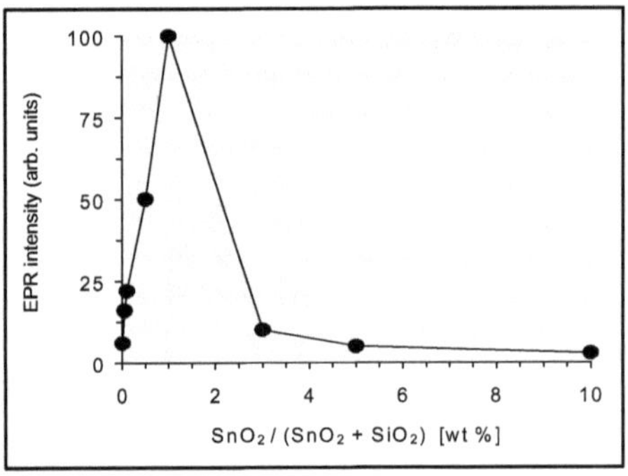

Fig. 4. (left) EPR spectrum at 298 K of *E*' Sn defect in Sn-doped silica glass (Sn 0.500 wt%); (right) EPR intensities of *E*' Sn signals vs. Sn content.

The substitutional position of Sn in SiO_2 tetrahedral sites was demonstrated by the presence of the paramagnetic E'-Sn centers, a three-coordinated tin center with an unpaired electron in a sp^3-like orbital, formed in Sn-doped silica glass by X-ray irradiation.[11] The signals of the *E*' Sn centers in orthorhombic symmetry field has g_1 = 1.994, g_2 = 1.986, g_3 = 1.975 whatever the amount of dopant (as an example EPR spectrum of 0.5 wt % sample is reported in Figure 4a). The EPR intensities of *E*' Sn centres increased with the amount of dopant in the colorless glasses but was significantly lower in yellow glasses where the presence of SnO_2 particles was observed (Figure 4b). The spectra of all samples also showed the signals of silicon-related defects in irradiated silica: the narrow and asymmetric line at about g = 2.001 attributed to *E*' Si centers [25] and the resonances attributed to oxygen related sites, non-bridging oxygen hole centers ≡Si-O• and peroxy radicals ≡Si-O-O•, at g values 2.002, 2.008 and 2.009.[26]

References

[1] K. O. Hill, G. Meltz, J. Lightwave Tech. 1997, **15**, 1263.
[2] G. Brambilla, V. Pruneri, L. Reekie, Appl. Phys. Lett. 2000, **76**, 807.
[3] B. Poumellec, F. Kherbouche, J. Phys. III, 1996, **6**, 1595.
[4] R. M. Atkins, V. Mizrahi, T. Erdogan, Electron. Lett. 1993, **29**, 385.
[5] M. G. Seats, G. R. Atkins, S. B. Poole, Annu. Rev. Mater. Sci. 1993, **23**, 381.
[6] T. E. Tsai, C. G. Askins, J. Friebele, Appl. Phys. Lett. 1992, **61**, 390.
[7] N. Chiodini, S. Ghidini, A. Paleari, Phys. Rev.B 2001, 64.
[8] L. Dong, J. L. Cruz, L. Reekie, M. G. Xu, D. N. Payne, IEEE Photon. Technol. Lett. 1995, **7**, 1048.
[9] K. Imamura, T. Nakai, Y. Sudo, Y. Imada, Electronics Lett. 1998, **34**, 1772.
[10] C. J. Evans, S. Karpel, Organotin Compounds in Modern Technology, J. Organomet Chem. Library 16, Elsevier, 1985.
[11] N. Chiodini, F. Meinardi, F. Morazzoni, A. Paleari, R. Scotti, G. Spinolo, Phys. Rev. B 1998, **58**, 9615.
[12] N. Chiodini, F. Morazzoni, A. Paleari, R. Scotti, G. Spinolo, J. Mater. Chem. 1999, **9**, 2653.
[13] N. Chiodini, F. Meinardi, F. Morazzoni, A. Paleari, R. Scotti, G. Spinolo, Solid State Commun. 1998, **109**, 145.
[14] C. Canevali, N.Chiodini, F.Morazzoni, J. Padovani, A. Paleari, R. Scotti, G. Spinolo, J. Non-Cryst. Solids 2001, **293-295**, 32.
[15] N. Chiodini, F. Meinardi, F. Morazzoni, J. Padovani, A. Paleari, R. Scotti, G. Spinolo, J. Mater. Chem. 2001, **11**, 926.
[16] J. Robertson, J. Phys. C 1979, **12**, 4767.
[17] A. Anedda, C. M. Carbonaro, A. Serpi, N. Chiodini, A. Paleari, R. Scotti, G. Spinolo, G. Brambilla, V. Pruneri, J. Non-Cryst. Solids 2001, **280**, 287.
[18] L. Skuja, J. Non-Cryst. Solids 1992, **149**, 77.
[19] L. Abello, B. Bochu, A. Gaskov, S. Koudryavtseva, G. Lucazeau, M. Roumyantseva, J. Solid State Chem. 1998, **135**, 78.
[20] F. L. Galeener, Solid State Commun. 1982, **44**, 1037.
[21] F. L. Galeener, Phys. Rev. B 1978, **19**, 4292.
[22] R. A. Murray, W. Y. Ching, Phys. Rev. B 1989, **39**, 1320.
[23] R. S. Katiyar, P. Dawson, M. M. Hargreave, G. R. Wilkinson, J. Phys. C: Solid State Phys. 1971, **4**, 2421.
[24] E. Geissberger, F. L. Galeener, Phys. Rev. B 1983, **28**, 3266.
[25] M. Stapelbroek, D. L. Griscom, E. J. Friebele, G. H. Sigel Jr., J. Non-Cryst. Solids 1979, **32**, 313.
[26] D. L. Griscom, Nucl. Instrum. Methods Phys.Res. B 1984, **1**, 481.

Octaphenyloctasilsesquioxane and Polyphenylsilsesquioxane for Nanocomposites

S.-G. Kim, S. Sulaiman, D. Fargier and R. M. Laine

Abstract A simple two step reaction was developed to prepare octaphenyloctasilsesquioxane ($[PhSiO_{1.5}]_8$, OPS) in high yield from phenyltrichlorosilane ($PhSiCl_3$). Octaphenyloctasilsesquioxane is easily modified by a wide number of electrophilic reactions to provide octa and hexadeca functionalized compounds that can be used as three-dimensional building blocks for nanometer-by-nanometer construction of composite materials. In this study, $PhSiCl_3$ was reacted with ethanol under reflux to produce two products, one is the liquid phenyltriethoxysilane [$PhSi(OEt)_3$, PTES], the other is an uncharacterized polymeric material likely, $EtO[PhSiO(OEt)]_n$. OPS was synthesized from both products. PTES gave primarily one crystal form of OPS. However, this material was contaminated by another compound not be easily separated because of the insolubility of OPS. In contrast, the polymeric version of PTES, when dissolved and reacted under identical conditions, gave a second phase of OPS analytically pure as formed. This same polymeric starting material, when reacted with catalytic amounts of KOH in ethanol, provided high molecular weight polyphenylsilsesquioxane (PPS) with only small amounts of OPS.

Classification

form:	molecular or polymeric solid
function:	precursor for inorganic-organic hybrid materials
preparation:	hydrolysis / condensation
composition:	$(C_6H_5SiO_{1.5})_n$

Introduction

Polyhedral oligosilsesquioxanes (POSS) are attractive compounds for numerous applications with structures derived from hydrolytic condensation of trifunctional organosilanes ($RSiX_3$, X = halogen, alkoxy, etc.). Since their discovery in 1946,[1] numerous studies have focused on the synthesis of POSS by hydrolysis and condensation of trifunctional silanes.[2-16] It is now generally recognized that high yield preparative routes to POSS materials are not always easily realized from $RSiX_3$ compounds because their hydrolysis and condensation kinetics and thermodynamics are strongly controlled by the nature of the substituent R and by the reaction conditions used. Thus, some POSS compounds are best synthesized under acidic conditions and others using basic conditions. Many studies have focused on the chemistry and properties of the resultant materials, particularly on the stoichiometrically well-defined POSS frameworks, including those with synthetically useful functional groups.[17-33] These materials are of considerable interest because of their unusual three-dimensional molecular architecture, their nanometer diameters, thermal stability and the extensive variety of functional groups that can be appended directly to the core. The core (0.5 nm body diagonal) is the smallest single crystal of silica.

Phenyl substituted POSS ($[PhSiO_{1.5}]_8$, OPS) represents one of the more interesting compounds because of its very high thermal stability.[34] Recently, it was shown, that OPS is easily octa-functionalized using common electrophilic reactions to produce materials that can be assembled nanometer by nanometer to give highly tailored materials.[35-37] However, the preparation of high quality OPS using published preparations is not easy. Until recently, OPS synthesis relied on methods reported by Barry,[7] Sprung,[4] Olsson,[9-10] and especially Brown et al.[12]

The first oligophenylsilsesquioxane, $[PhSiO_{1.5}]_6$, was obtained as a crystalline precipitate by Barry et al in 1955.[7] Sprung and Guenther thereafter reported that $[PhSiO_{1.5}]_8$ formed by rearrangement of a high polymer $[PhSiO_{1.5}]_n$ produced from a mixture of phenyltriethoxysilane, water, tetraethylammonium hydroxide, and methyl isobutyl ketone.[4] Olsson et al. [9-10] and Brown et al.[12] prepared $[PhSiO_{1.5}]_{8-12}$ by base-catalyzed equilibration of polyphenylsilsesquioxane at reflux.

Unfortunately, these reported methods are inconvenient because they are multistep procedures that require large quantities of starting materials, involve toxic reactants and long reaction times and provide low yields of the desired products. Because OPS is poorly soluble in almost all solvents and decomposes before it melts there was little incentive to improve on the published syntheses. With our discovery that OPS is amenable to electrophilic functionalization and that the functionalized materials provide access to novel nanocomposites with highly tailored properties, there is now renewed interest in producing large quantities of OPS using simple methods. We herewith describe a simple, easily scaled method of producing pure OPS as well as a simple route to a high molecular weight polymeric equivalent (Scheme 1).

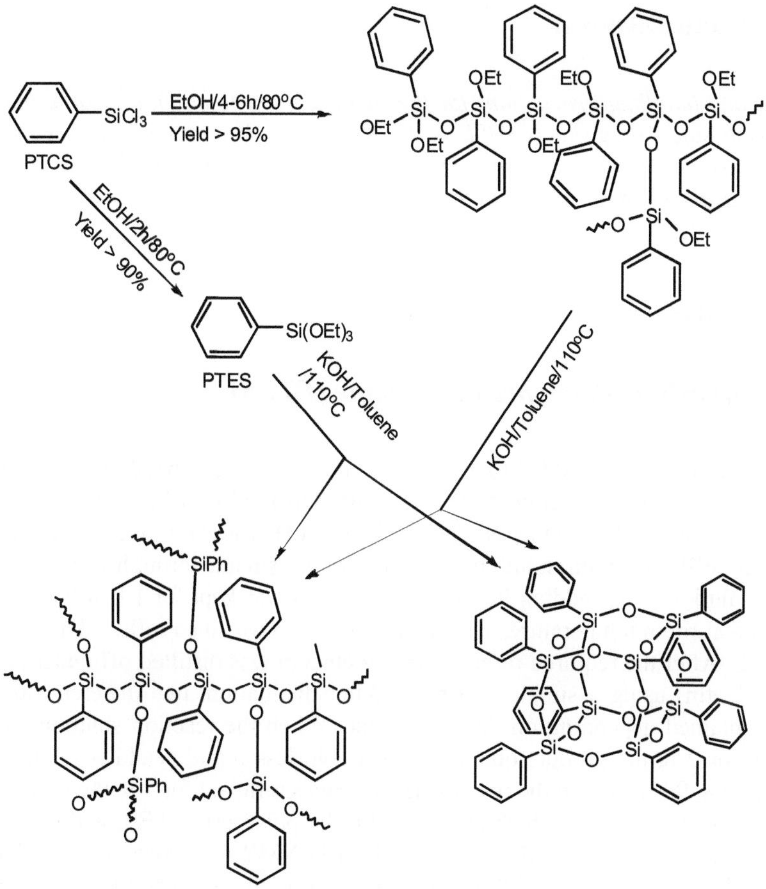

Scheme 1. General pathways for formation of OPS and PPS.

Materials

- Phenyltrichlorosilane ($PhSiCl_3$) was obtained from Aldrich or Clariant Life Sciences and distilled under N_2 before use.
- Tetrahydrofuran (THF) and toluene were obtained from Fisher Scientific and freshly distilled from sodium/benzophenone ketyl before use.
- Anhydrous ethanol, methanol, and hexane were obtained from Fisher Scientific and used without purification.
- Potassium hydroxide (KOH) was obtained from Aldrich and used without purification.

Safety and Disposal

Safety and handling instructions for the chemicals are found in the corresponding materials safety data sheets (MSDS). All products and byproducts are used in subsequent reactions. Waste produced comprises organic solvents that can be recycled or disposed of according to their MSDS's.

Procedures

A. Synthesis of Phenyltriethoxysilane (PTES)

Ethanol (28 mL) was added to a 250 ml Schlenk flask equipped with a magnetic stir bar, a cooling condenser, and an ice bath. Distilled $PhSiCl_3$ (7 g, 33.1 mmol) was then slowly added under N_2. The by-product HCl was removed by passing N_2 just above the reaction solution and venting this stream through a concentrated KOH solution. After adding $PhSiCl_3$, the solution was kept for 1 h at 0 °C, stirred for 20 h at room temperature. The reaction solution was then refluxed at 80 °C for 2 or 4 h. After the required reflux time, the ethanol was distilled off under N_2 (80-90 °C/1 atm) using a simple still head. After distillation, two different products were obtained depending on the reflux time. When the reaction solution was refluxed for 2 h, the product obtained was a colorless liquid (yield = 7.1 g PTES, 29.5 mmol, 90%). When the reaction mixture was refluxed for 4 h, the product obtained was a white gel, which is assumed to be polymeric PTES, with a yield of 6.9 g (28.7 mmol, 87%). <u>Monomeric PTES</u> (^{1}H NMR 400 MHz, acetone-d_6): 7.6 (Ar, 2.0H), 7.4 (Ar, 3.0H), 3.9 (6.0H), 1.3 (9.0H) ppm. <u>Polymeric PTES</u> (^{1}H NMR 400 MHz, acetone-d_6): 7.9-7.1 (Ar, 6.5H), 4.2-3.3 (2.0H), 1.4-0.7 (3.0H) ppm.

B. Synthesis of OPS and its Polymeric Analog PPS from PTES.

Various solvents were examined in an effort to produce OPS from PTES using KOH as a catalyst. Toluene, ethanol, and mixtures of toluene/ethanol were used.

Synthesis of OPS and PPS from monomeric PTES. Monomeric PTES (7 g, 29.1 mmol), synthesized as above, was added to a 100 ml round-bottomed flask equipped with a magnetic stir bar and a cooling condenser. Solvent (50 ml) and potassium hydroxide (0.04-0.75 g, 0.5-10 wt% based on PTES) was then added under N_2. The solution was heated to reflux at 110 °C, and then water (0.125-1.5 ml, 2.5-30 wt% based on PTES) was slowly added in small portions of 0.1-0.2 ml each 3-5 min over a 30 min. The reactions were kept refluxing for 6-80 h. After about 3 h, a white powder (OPS) begins to precipitate. After the required time, the precipitated powder is filtered off and washed with methanol (3 x 20 ml). The clear filtrate was distilled (120 °C/1 atm) to remove residual toluene and water, and concentrated to 20-30 mL. The remaining liquid was then slowly precipitated into hexane (100 ml). The resulting precipitated product (PPS) was filtered and washed with hexane (3 x 20 mL).

Both products are vacuum dried at 70 °C/7 h. The products obtained were a mixture of OPS and a polymeric form (PPS) (soluble in methanol) in varying yields (Table 1). The maximum yield for OPS was 3.41 g (yield 91 %) from toluene when 0.5 ml water (10 wt% water based on PTES) was used and the reaction run for 60 h. The maximum yield of the polymeric version was 3.32 g (yield 88 %) from ethanol when 0.5 ml water (10 wt% water based on PTES) was used and the reaction run for 20 h. The resulting materials were characterized by XRD, ^{29}Si NMR etc (Table 1).

Synthesis of OPS and PPS from polymeric PTES. Polymeric PTES, 7 g (29.1 mmol), synthesized as above, was added to a 100 mL round-bottomed flask equipped with a magnetic stir bar and a cooling condenser under flowing N_2. Toluene (50 mL) and KOH (0.5 g, 7.5 wt% based on PTES) were then added. The solution was heated to reflux at 110°C, and then water (0.5 ml, 27.8 mmol, 10 wt% based on PTES) was added slowly as noted above. Reactions were refluxed from 6-80 h to examine the effect of time with optimum yields obtained at 60 h for OPS (toluene) and at 20 h for PPS (ethanol). White OPS powder slowly precipitated and was recovered by filtration. The clear filtrate was treated as above to obtain PPS. The OPS powder, treated as above, was characterized by mass spectrometry, XRD, ^{29}Si NMR etc (Table 2). The maximum yield of OPS was 3.43 g (3.3 mmol, 91 %). Remaining, soluble PPS was recovered by precipitation giving 1.05 g (1.01 mmol, 28 %).

Characterization

MALDI-TOF mass spectrometry (Table 1): Matrix-assisted ultraviolet laser desorption/ionization time-of-flight mass spectrometry (UV-MALDI-TOF MS) was performed using a TOF SPEC-2E/MALDI™ (Micromass, Inc,) equipped with a pulsed nitrogen laser (λ = 337 nm, pulse width = 3 ns, average power at 20 Hz = 5mW). The extraction voltage in TOF analyzer was 20 kV, and ions were obtained by irradiation just above the threshold laser power. The measurement was carried out by applying of matrix and sample solution on the graphite plate.

Solutions for analysis prepared from very dilute solutions of OPS were too low in concentration. Therefore, a dispersion prepared in methylene chloride was used without filtering, and provided the correct mass peaks. For benchmarking purposes we compared our products with OPS purchased from Hybrid-Plastic. OPS obtained from monomeric and polymeric PTES gives a formula weight of 1095, as does the purchased OPS.

Table 1. Fragmentation pattern from the MALDI

OPS*	OPS	PPS	DDPS *
1095.2 (highest)	1095.3 (middle)	1095.2 (highest)	1613.4 (highest)
955.2	1015.4	955.2	1472.4
879.4	955.2 (highest)	879.5	1400.4
832.4	879.5	698.0	1324.7
682.0	817.4	682.0	1248.5

*As-received from Hybrid Plastic. DDPS = dodecaphenylsilsesquioxane, $[PhSiO_{1.5}]_8$.

NMR spectroscopy: All CP-MAS solid-state NMR spectra were obtained at 9 T using a Chemagnetics CMX-400 spectrometer operating at 79.5 MHz for ^{29}Si and 400.13 MHz for ^{1}H. Contact times were 2 ms with 20 s pulse delays. The probe used was a Chemagnetics PENCIL design using 5 mm zirconia rotors at spinning rates of 3 kHz. The sample temperature was at 30 °C and TMS referenced the chemical shifts.

The nature of the T units in POSS, silane diol (T_1), silanol (T_2), siloxane (T_3), were identified by their chemical shifts using ^{29}Si-CP-MAS NMR spectra. The half-height full width (HHFW) values of the T peaks can be used as parameters for structural analysis.[31-32] ^{29}Si-CP-MAS NMR spectra of products obtained by refluxing monomeric and polymeric PTES in toluene are shown in Fig. 1.

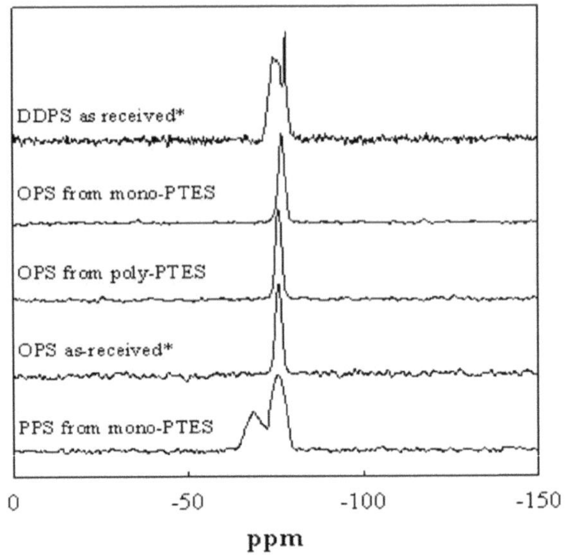

Fig. 1. ^{29}Si-CP-MAS solid-state NMR spectra of DDPS, PPS and OPS. *from Hybrid-Plastic Co.

The isolated products give sharp singlets at -75.9 and -76.1 ppm (T_3), respectively. Within the error limits, these are considered to be the same. The recovered soluble product exhibits two broad doublets at -69.3 ppm (T_2) and -76.4 ppm (T_3). Thus insoluble products consist solely of T_3 units, whereas the soluble PPS appears to consist of typical T units and bridging silanol groups as suggested by Fig. 1 and based on work by Feher et al.[19-20] That is, the white powders precipitated during reaction have perfect cage structures and the soluble products consist of polymers derived from broken-cage structures.

Frye and Collins [38] reported that ^{29}Si NMR spectra of octa- and deca(hydridosilsesquioxane) display singlets, whereas dodeca(hydridosilsesquioxane) displays two singlets of different intensities consistent with structural isomers. Likewise, as-prepared OPS and the commercial sample have a singlet at -75.9 ppm whereas the commercial dodecaphenyl DDPS material has two singlets at -75.0 and -78.0 ppm. Thus, the ^{29}Si NMR data ensure that the products generated here are not DDPS but OPS. Table 2 provides the peak positions and HHFW for products. For the cage compounds, the HHFW are very small compared to the PPS values.

Table 2. ^{29}Si NMR spectra and HHFW for OPS and PPS

Starting material	Product	δ (ppm)	HHFW (ppm)
	OPS*	-75.9	1.8
	DDPS*	-75.0 / -78.0	5.0 / 1.2
$PhSi(OEt)_3$	OPS	-76.1	1.8
	PPS	-69.5 / -76.2	6.3 / 5.1
Polymeric PTES	OPS	-75.9	1.8
	PPS	-69.3 / -76.4	6.0 / 5.0

*Purchased or received as a gift from Hybrid Plastics.

Fourier transform infrared spectroscopy (FTIR): Spectra were recorded on a Mattson Galaxy Series 3020 bench adapted with a Harrick Scientific "Praying Mantis" DR accessory (DRA-2CO). KBr was used as a nonabsorbent medium. Sample was ground with KBr to make a 1 wt % mixture and packed tightly in the sample holder. After the sample was loaded into the chamber, nitrogen was purged for about 10 min before data collection. A minimum of 32 scans was collected for each sample at a resolution of 4 cm^{-1}.

Fig. 2 shows FTIR spectra for the products obtained from monomeric and polymeric PTES. There is little difference in the spectra. The FTIR spectra are all characterized by two broad maxima associated with νSi-O-Si absorptions in the 1200-950 cm^{-1} region. Brown et al report that $[PhSiO_{1.5}]_{8-12}$ exhibit only one $\nu_{Si\text{-}O\text{-}Si}$ band at 1120-1130 cm^{-1}, while PPS exhibits two bands at 1135-1150 and 1045-1060 cm^{-1}.[12] As shown in Fig. 2, PPS generates two broad $\nu_{Si\text{-}O\text{-}Si}$ bands centered at 1126 and 1050 cm^{-1}, whereas OPS exhibits only one sharp singlet peak at 1124 cm^{-1}. This also suggests that PPS consists of cage and ladder structures.

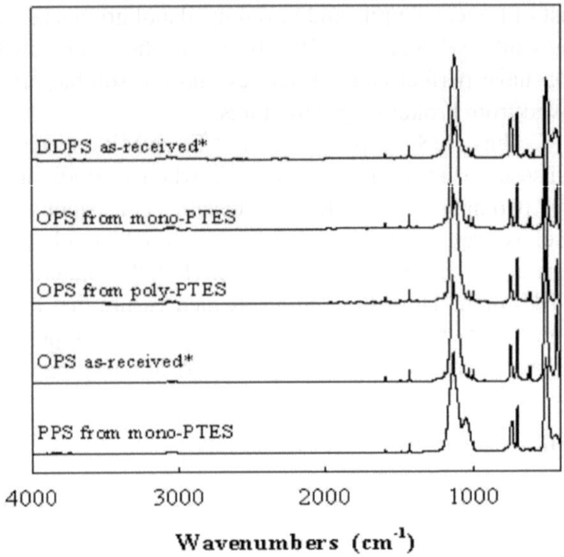

Fig. 2. FTIR spectra of commercial DDPS*and OPS*, OPS from polymeric PTES, OPS from PTES, and PPS from PTES.

The spectra of octa- and dodecasilsesquioxanes are also characterized by three or four intense bands in the 360-600 cm^{-1} region arising from symmetric deformational vibrations of the silicon-oxygen (Si-O-Si) framework. Per Fig. 3, OPS exhibits slightly different and sharper peak positions and shapes compared to DDPS because of structural isomers.

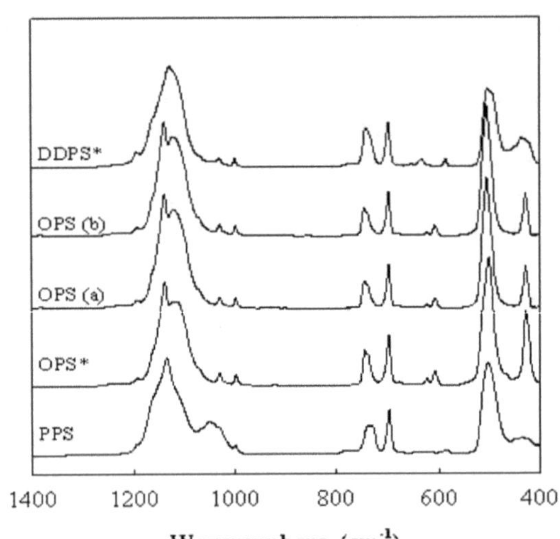

Fig. 3. FTIR spectra of commercial DDPS*and OPS*, OPS[a] from polymeric PTES, OPS[b] from monomeric PTES, and PPS from monomeric PTES.

Thermal gravimetric analyses (TGA) were performed on a SDT 2960 simultaneous DTA-TGA thermogravimetric analyzer (TA instrument, Inc., New Castle, DE). The instrument was calibrated with Alumel and iron supplied by TA. Measurements were performed under a continuous flow of synthetic air (110 ml/min.), at 10 °C/min to 1000 °C.

TGAs for OPS and PPS obtained from PTES and polymeric PTES all have similar shapes and decomposition points over 500 °C in the air. OPS has previously been reported to offer excellent thermal stability as has PPS (> 500 °C). The thermal stability of decaphenylsilsesquioxane ($[PhSiO_{1.5}]_{10}$) and DDPS are 415–418 and 385 °C, respectively.[10,13,39] The relative stability of silsesquioxanes with n = 6, 8, 10, 12, 14 is generally presumed to be determined mainly by the degree of distortion of the Si-O-Si angle. It was also reported that the decomposition points of $(CH_3SiO_{1.5})_n$ with 6, 8, 10, 12 are 209-210, 415, 333-334, and 270 °C, respectively.[34-36]

Gel permeation chromatography (GPC) analyses were performed on a Waters GPC system, using a Waters 410 RI detector and a Waters 486 UV detector, Waters Styragel columns (7.8 x 300, HR 0.5, 1, 3, 4), and a PL-DCU data capture unit from Polymer Laboratory. The system was calibrated using polystyrene standards obtained from Polymer Laboratory. THF was used as the eluent, at a flow rate of 1.0 ml/min.

The M_n and M_w of OPS obtained from polymeric PTES are 699 and 701, respectively, giving a polydispersity of 1.00. OPS obtained from PTES is essentially identical. The M_n, M_w, and polydispersity of PPS obtained from the filtrate are 3610, 6890, and 1.9, respectively.

X-ray diffraction (XRD) analysis were run using a Rigaku Rotating Anode Goniometer (Rigaku Denki Co. Ltd., Tokyo, Japan). The working voltage and current were 49 kV and 100 mA respectively. Cu *K*α (λ = 1.54 Å) radiation with a Ni filter was used. Powder was mounted and pressed on a glass holder and scanned from 2° to 40° in increments of 0.2°. Bragg's law was used to calculate the d spacings.

Fig. 4 shows the XRD analyses of the products obtained from monomeric and polymeric PTES. PPS exhibits only amorphous scattering, but the products derived from monomeric and polymeric PTES show sharp peaks. It appears that the products from the two different starting materials exhibit different crystal morphologies. Larsson and Olsson et al. observed two crystalline morphologies for OPS; one is triclinic with one molecule per unit cell, the other has a monoclinic unit cell containing two molecules.[10,36] Brown et al characterized both crystal structures,[12] by XRD. The characteristic powder patterns for both morphologies I and II are given in Table 3.

The cage product from polymeric PTES can be assigned to OPS of Form I, whereas the cage product from monomeric PTES appears to be Form II according to observed lattice parameters from XRD measurements as compared with the published values.[12] The OPS from polymeric PTES matches the commercial sample identically (Table 3). These results agree with the MALDI-TOF data.

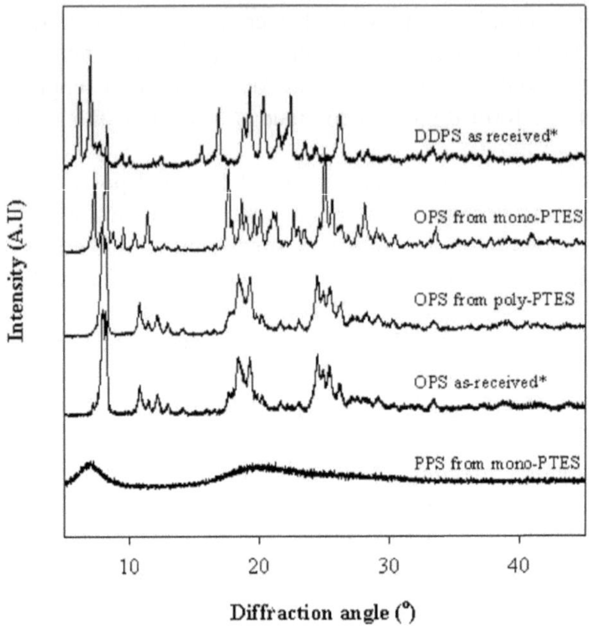

Fig. 4. XRD peaks of PPS and OPS obtained from both forms of PTES, water, and KOH in toluene. *from Hybrid-Plastic Co.

Table 3. d-spacings (Å) for most prominent reflections of OPS and PPS obtained from PTES and polymeric PTES and various reported values (Forms I and II)[12]

OPS*		DDPS*		OPS†	DDPS†	OPS‡	
I	II	I	II	-	-	Mono-PTES	Poly-PTES
10.9	12.0	12.3	13.0	10.9	14.0	12.0	10.9
8.2	10.6	11.8	12.0	8.1	12.4	10.7	8.2
7.3	10.1	11.1	11.4	7.3	9.5	10.1	7.3
4.8	9.4	9.3	10.6	4.8	5.2	9.5	4.8
4.6	8.4	-	10.1	4.6	4.6	8.5	4.6
3.6	7.7	-	8.4	3.6	3.4	7.7	3.6
-	3.9	-	-	-	-	3.8	-

*d-spacing values of cage compounds reported by Brown et al.[12] †d-spacing values of OPS and DDPS obtained from Hybrid Plastics Co. ‡d-spacing values of cage compounds prepared by our method

Comments

Previous methods of forming OPS were complex, multi-step and low-yield processes. Above we described a simple, two-step route to high yields of $[PhSiO_{1.5}]_n$

per Scheme 1. The process begins with the conversion of $PhSiCl_3$ into PTES by reaction in ethanol. Surprisingly, depending on the reaction conditions (reflux for 2 h vs. 4-6 h), the product is either the liquid monomer PTES or a polymeric version (polymeric PTES), both of which provide OPS with some polymeric material $[PhSiO_{1.5}]_n$, PPS by reaction with catalytic amounts of KOH and minimal amounts of water. Both products are easily separated and purified.

As noted above, a wide variety of reaction conditions have been used to identify superior conditions for obtaining high yields of POSS from $RSiX_3$. The exact choice of conditions is predicated on the R group and the X group. OPS produced by hydrolysis of $PhSiCl_3$, forms more readily in benzene, nitrobenzene, benzyl alcohol, pyridine, ethylene glycol dimethyl ether, whereas the dodecamer ($[PhSiO_{1.5}]_{12}$, DDPS) is formed in tetrahydrofuran. Hydrolysis of $PhSiCl_3$ in acetonitrile, diglyme, acetone, and methyl isobutyl ketone gives high MW PPS.[12] Most OPS syntheses use benzene as a solvent which is now known to be carcinogenic, thus toluene was used as a substitute.[7,10,12]

Water scarce conditions are the key to the successful synthesis of OPS. Water must be added carefully dropwise over about 30 min. In our standard reaction, 0.125-1.5 ml (6.9-83.3 mmol) of water is added slowly to solutions containing 7 g PTES (29.1 mmol) and 50 ml solvent. The concentration of water to be added must be calculated exactly to ensure optimal yield, 27.8 mmol (29.1 theory), of OPS. *The concentration of water should be just slightly less than the equivalents of PTES or Polymeric PTES used.* We assume that some adventitious water, present in the reaction system, makes up the difference in stoichiometries used. Polymeric PTES always gives better yields of OPS than PTES probably because some of the cage structure is preformed in the polymer (Fig. 5). The yield of OPS and PPS are affected by concentration of KOH, water, reaction time, and solvents.

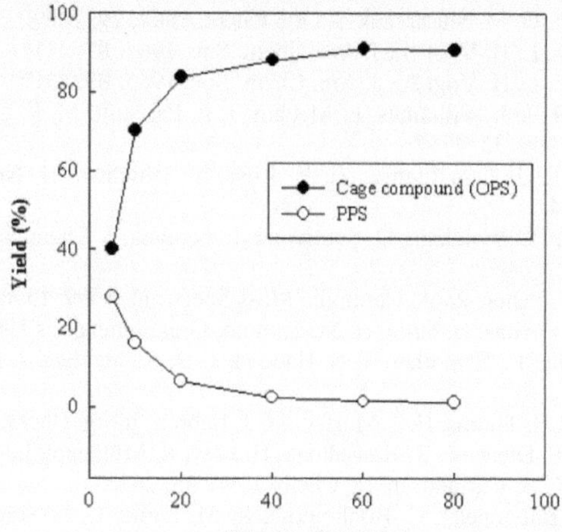

Fig. 5. Yield of OPS and PPS formed by reflux with polymeric PTES gel in toluene under selected conditions.

OPS was prepared in yields >85 % in 20 h, and over 90% in 60 h in toluene. When pure ethanol is used as solvent with 10 wt% water and 7.5 wt% KOH (based on PTES), little OPS forms and high MW PPS can also be obtained in yields >88 % after 20 h. The M_n, M_w, and polydispersity of the PPS are $1.6 \cdot 10^4$, $2.7 \cdot 10^4$, and 1.7, respectively. In toluene, the PPS formed has low MWs (M_n: about $2.5\text{-}4.0 \cdot 10^3$, M_w: $3.5\text{-}6.5 \cdot 10^3$). This process differs from previously reported methods [7-10] and is a very simple and economic synthesis.

OPS is highly insoluble making detailed characterization quite difficult. However, it is somewhat more soluble in methylene chloride and pyridine than other solvents, thus most solution spectroscopy is best done in these solvents.

References

[1] D. W. Scott, J. Am. Chem. Soc. 1946, **68**, 356.
[2] M. M. Sprung, F. O. Guenther, J. Am. Chem. Soc. 1955, **77**, 3990.
[3] M. M. Sprung, F. O. Guenther, J. Am. Chem. Soc. 1955, **77**, 3996.
[4] M. M. Sprung, F. O. Guenther, J. Polymer. Sci. 1958, **28**, 17.
[5] M. M. Sprung, F. O. Guenther, J. Am. Chem. Soc. 1955, **77**, 6045.
[6] R. West, J. Am. Chem. Soc. 1953, **75**, 1002.
[7] A. J. Barry, W. H. Daudt, J. J. Domiconr, J. W. Gilrey, J. Am. Chem. Soc. 1955, **77**, 4248.
[8] J. F. Brown Jr., L. H. Vogt Jr., J. R. Katchman, J. W. Eustance, K. M. Kiser, K. W. Krantz, J. Am. Chem. Soc. 1960, **82**, 6194.
[9] K. Olsson, Arkiv Kemi. 1958, **13**, 367.
[10] K. Olsson, C. Gronwall, Arkiv Kemi. 1961, **17**, 529.
[11] L. H. Vogt Jr., J. F. Brown Jr., Inorg. Chem. 1963, **2**, 189.
[12] J. F. Brown Jr., L. H. Vogt Jr., P. I. Prescott, J. Am. Chem. Soc. 1964, **86**, 1120.
[13] J. F. Brown Jr., P. I. Prescott, J. Am. Chem. Soc. 1964, **86**, 1402.
[14] J. F. Brown Jr., G. M. Slusarczuk, J. Org. Chem. 1964, **29**, 2809.
[15] J. F. Brown Jr., L. H. Vogt Jr., J. Am. Chem. Soc. 1965, **87**, 4313.
[16] J. F. Brown Jr., L. H. Vogt Jr., J. Am. Chem. Soc. 1965, **87**, 4317.
[17] D. P. Fasce, R. J. J. Williames, F. Mechin, J. P. Pascault, M. F. Llauro, R. Petiaud, Macromolecules 1999, **32**, 4757.
[18] D. P. Fasce, R. J. J. Williames, E. B. Rosa, Y. Ishikawa, H. Nonami, Macromolecules 2001, **34**, 3534.
[19] F. J. Feher, K. D. Wyndham, D. Soulivong, F. Nguyen, J. Chem. Soc., Dalton Trans. 1999, 1491.
[20] R. Bakhtiar, F. Feher, Rapid Commun. Mass Spectrom. 1999, **13**, 687.
[21] M. Unno, S. B. Alias, H. Satio, H. Matsumoto, Organometallics 1996, **15**, 2413.
[22] A. Romo-Uribe, P. T. Mather, T. S. Haddad. J. D. Lichtenhan, J. Polym. Sci. B 1998, **36**, 1857.
[23] B. Hong, T. P. S. Thoms, H. J. Murfee, M. J. Lebrun, Inorg. Chem. 1997, **36**, 6146.
[24] R. Knischka, F. Dietsche, R. Hanselman, H. Frey, R. Mülhaupt, Langmuir 1999, **15**, 4752.
[25] S. E. Yuchs, K. A. Carrado, Inorg. Chem. 1996, **35**, 261.
[26] C. Zhang, F. Babonneau, C. Bonhomme, R. M. Laine, C. L. Soles, H. A. Hristov, A. F. Yee, J. Am. Chem. Soc. 1998, **120**, 8380.
[27] F. J. Feher, D. Soulivong, G. T. Lewis, J. Am. Chem. Soc. 1997, **119**, 11323.
[28] F. J. Feher, J. J. Schwab, D. Soulivong, J. W. Ziller, Main Group Chem. 1997, **2**, 123.
[29] C. Zhang, R. M. Laine, J. Am. Chem. Soc. 2000, **122**, 6979.
[30] P. A. Agaskar, Inorg. Chem. 1991, **30**, 2707.

[31] L. A. S. D. A. Prado, E. Radovanovic, H. O. Pastore, I. V. P. Yoshida, I. L. Torriani, J. Polym. Sci. A 2000, **38**, 1580.
[32] E. C. Lee, Y. Kimura, Polymer J. 1998, **30**, 234.
[33] E. C. Lee, Y. Kimura, Y. Polymer J. 1998, **30**, 730.
[34] R. H. Baney, M. Itoh, A. Sakakibara, T. Suzuki, Chem. Rev. 1995, **95**, 1409.
[35] R. Tamaki, Y. Tanaka, M. Z. Asuncion, J. Choi, R. M. Laine, J. Am. Chem. Soc. 2001, **123**, 12416.
[36] C. M. Brick, Y. Ouchi, Y. Chujo, R. M. Laine, Macromolecules 2005, **38**, 4661.
[37] C. M. Brick, R. Tamaki, S.-G. Kim, M. Z. Asuncion, M. Roll, T. Nemoto, R. M. Laine, Macromolecules 2005, **38**, 4655.
[38] C. L. Frye, J. M. Klosowski, J. Am. Chem. Soc. 1971, **93**, 4599.
[39] K. Larsson, Arkiv Kemi. 1960, **16**, 209.

Polysilsesquicarbodiimide Xerogels

S. Nahar-Borchert, A. O. Gabriel and R. Riedel

Abstract Polysilsesquicarbodiimide xerogels have been prepared by sol–gel processing of chlorosilanes, R_xSiCl_{4-x} (R = H, alkyl or aryl; x = 0, 1, or, 2) and bis(trimethylsilyl)carbodiimide. The reactions can be performed with or without organic solvents and are catalyzed by pyridine. Depending on the solvents and the applied experimental conditions, the product is obtained as a fine powder or as a gel. Heat treatment transforms the xerogels to ceramic materials in the ternary Si/C/N system.

Classification

form:	amorphous monoliths, powder
function:	ceramic precursor
preparation:	non-oxidic sol–gel processing
composition :	organically modified silicon carbonitride SiCN

Introduction

Advanced non-oxide ceramics and ceramic composites with compositions in the ternary Si–C–N system are of high technical relevance.[1] Polysilsesquicarbodiimides have been successfully applied as single-source precursors for the synthesis of novel ternary Si-, C-, and N-containing solid phases. Their thermally induced decomposition gives either amorphous silicon carbonitrides or polycrystalline silicon nitride and silicon carbide mixtures. These materials are presently of technological interest for their exceptional hardness, strength, toughness, and high temperature resistance even in corrosive environments.[2-4] Additionally, the reactive carbodiimide group provides easy introduction of, for example, boron.[5-7] These boron containing precursors can be pyrolyzed to Si–B–C–N ceramics, which exhibit high thermal stability.[8-9]

The described procedure allows the synthesis of poly(methylsilsesquicarbodiimide) xerogel, $[MeSi(N{=}C{=}N)_{1.5}]_n$ [10-11] from a reaction between stoichiometric amounts of methyltrichlorosilane and bis(trimethylsilyl)carbodiimide (Eq. 1). The reaction is performed without solvent and is catalyzed by pyridine. The gelation time can be controlled by varying the amount of catalyst and by changing the reaction temperature. By determination of a rate constant for the reaction (Eq. 1) that is dependent on the reciprocal temperature (Arrhenius plot, Fig. 1), the reaction time until gelation can be predicted. Each of the three lines in Fig. 1 represents a fixed pyridine proportion. The slopes of 0.8, 0.4, and 0.2 eq. (equivalents of pyridine with respect to the silane) have nearly the same value.

$$n\ MeSiCl_3 + 1.5n\ Me_3Si\text{-}N{=}C{=}N\text{-}SiMe_3 \xrightarrow{\text{pyridine}} \underbrace{[MeSi(N{=}C{=}N)_{1.5}]_n + 3n\ Me_3SiCl}_{\text{Gel}} \qquad (1)$$

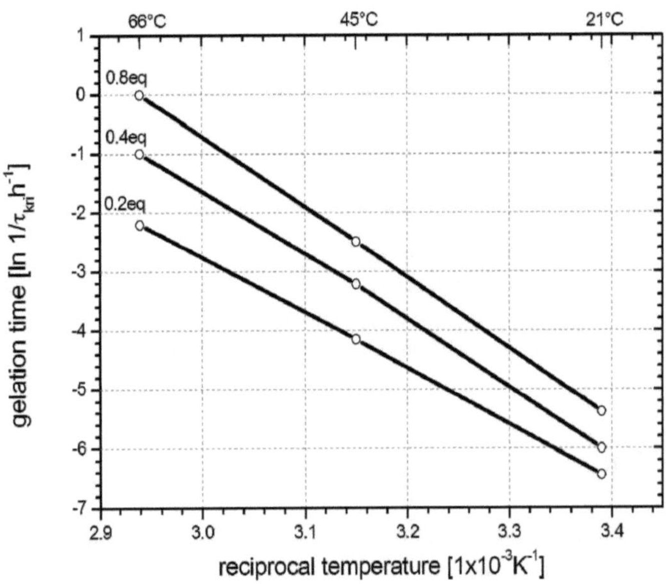

Fig. 1. Arrhenius plot for the synthesis of poly(methylsilylcarbodiimide) gels for three different pyridine equivalents (eq). The reciprocal gelation time τ is described in relation to the reciprocal temperature.

The synthesis of poly(methylsilsesquicarbodiimide) or, more generally, the reaction sequence for production of silylcarbodiimide polymers is closely related to that of the reaction of chlorosilanes, R_xSiCl_{4-x} (R = H, alkyl or aryl; $x = 0$, 1, or 2), with water, forming silica gels, silicones, or organically modified silicate materials. In the non-oxidic sol–gel process, bis(trimethylsilyl)carbodiimide adopts the role of H_2O applied in the conventional oxidic sol–gel route.[11]

The driving force of the reaction (Eq. 1) is the formation of stable Me_3SiCl, which can be easily separated from the polymeric carbodiimide gel by distillation. For liquid chlorosilanes the reaction can be carried out without a solvent.

The as-prepared gels are highly transparent. After an aging period of ~5 days at 45°C, the gel becomes cloudy without change of the gel volume. Further annealing at 45°C induces irreversible shrinkage. Careful evaporation of the liquid fraction, mainly Me_3SiCl, provides a transparent xerogel. The corresponding length and volume shrinkage values of the $[MeSi(NCN)_{1.5}]_n$ gel are summarized in Table 1.

Table 1. Relative length and volume shrinkage ($\Delta L/L$ and $\Delta V/V$) during aging and drying of the $[MeSi(NCN)_{1.5}]_n$ gel.

Dimensions	Gel as prepared	Aged gel[a]	Xerogel[b]
$\Delta L/L$ (%)	0	37	44
$\Delta V/V$ (%)	0	75	82

[a] Aging conditions: 30 d at 45°C; pyridine content 0.6 eq.
[b] Drying conditions: evaporation of the liquid phase at 50°C/1 bar for 24 h.

The rate of gel shrinkage is mainly determined by two parameters: increased pyridine content (0.0–1.5 eq) and prolonged annealing (20–45°C) accelerate the aging and consequently enhance the degree of cross-linking of the gel network.

Mechanism of Gel Formation

The formation of poly(methylsilsesquicarbodiimide) gel was explained by substitution of the chlorine atoms followed by condensation reactions.

Substitution

$MeSiCl_3 + Me_3Si\text{-}N{=}C{=}N\text{-}SiMe_3 \rightarrow MeCl_2Si\text{-}N{=}C{=}N\text{-}SiMe_3 + Me_3SiCl$

Condensation

$MeCl_2Si\text{-}N{=}C{=}N\text{-}SiMe_3 + ClSi{\equiv} \rightarrow MeCl_2Si\text{-}N{=}C{=}N\text{-}Si{\equiv} + Me_3SiCl$

$2\ MeCl_2Si\text{-}N{=}C{=}N\text{-}SiMe_3 \rightarrow MeCl_2Si\text{-}N{=}C{=}N\text{-}SiCl_2Me + Me_3Si\text{-}N{=}C{=}N\text{-}SiMe_3$

Materials

- $MeSiCl_3$ purchased from Aldrich and distilled under argon or nitrogen prior to use.

- *N, N'*-bis(trimethylsilyl)carbodiimide was synthesized according to the literature procedure.[11]

Safety and Disposal

Safety and handling instructions for the chemicals are found in the corresponding materials safety data sheets (MSDS). Methyltrichlorosilane and trimethylchlorosilane can cause severe skin and eye burns. All manipulations should be carried out in a well-ventilated fume hood; protective gloves and safety glasses should be worn.

Procedure

There are several ways to prepare bis(trimethylsilyl)carbodiimide.[12-14] However, from an economic point of view, the reaction of hexamethyldisilazane with cyanoguanidine, the dimer of cyanamide, is the most efficient route (Eq. 2).[11]

$$2\ [(CH_3)_3Si]_2NH + H_2NC(NH)NHCN \longrightarrow$$
$$2\ (CH_3)_3Si\text{-}N{=}C{=}N\text{-}Si(CH_3)_3\ + 2\ NH_3 \qquad (2)$$

All manipulations and syntheses were performed under purified argon or nitrogen atmosphere using standard Schlenk techniques.

Methyltrichlorosilane, $MeSiCl_3$ (10.7 g, 71.6 mmol), was mixed with bis(trimethylsilyl)carbodiimide (20.0 g, 107 mmol) and a catalytic amount of pyridine (1.70 g, 21.5 mmol) in a 100 cm^3 round-bottomed flask. The reaction mixture was heated to 90°C with stirring. The time until gelation depended strongly on the amount of pyridine. In the case of 0.3 eq. of pyridine (relative to $MeSiCl_3$), the time until gelation was about 4 h. After cooling to room temperature, the gel was aged by annealing at 45°C. The three-dimensional shrinkage of the gel was complete after about 50 d under these conditions. The xerogel, $[MeSi(NCN)_{1.5}]_n$ was obtained in 89% yield after evaporation of the volatiles (Me_3SiCl and residual bis(trimethylsilyl)carbodiimide) at room temperature and 50°C at 3×10^{-2} mbar).

Characterization

$[MeSi(NCN)_{1.5}]_n$ was obtained as a colorless, jelly-like solid.
IR (KBr pellets): 2965 [ν_{as}(C-H)], 2152 [ν_{as}(N=C=N)], 1270 [δ_s($SiCH_3$)], 796 [δ(N=C=N)], 566 [ν_{as}(Si-NCN)] cm^{-1}.

Raman (powder): 2974 and 2906 [ν_{as}(C-H)], 1533 [ν_s (N=C=N)] cm^{-1}
^{29}Si CP/MAS-NMR: δ = -62.3 ppm.
Elemental analysis: $C_{2.5}H_3N_3Si$ (103.16): Calcd. C, 29.11; N, 40.73; O, 0.00; Si, 27.23. Found: C, 28.75; N, 38.3; O, 0.28; Si, 26.9.

Upon exposure to air, $[MeSi(NCN)_{1.5}]_n$ is hydrolyzed to a mixture of cyanamide and silanols which eventually condense to polymeric siloxanes, $[\equiv Si\text{-}O\text{-}Si\equiv]_n$. The solid state FTIR spectrum contains the characteristic frequencies for the hydrolyzed products: 3272 [$\nu_{SiO\text{-}H}$ and $\nu_{N\text{-}H}$], 2264 [$\nu_{N\text{-}CN}$], 1578 [$\nu_{C\text{-}N}$], and 1063 [$\nu_{Si\text{-}O}$] cm^{-1}.

$[MeSi(NCN)_{1.5}]_n$ transforms to an amorphous silicon carbonitride ceramic, $SiC_{1.1}N_{1.6}$, by the thermally induced ceramization at 1200°C (holding time, 30 min) in an inert atmosphere (argon). The gel-derived silicon carbonitride is thermally stable up to 1450°C. At higher temperatures, pure crystalline β-SiC is formed ($SiC_{0.96}N_{0.04}$).

Comments

This method has been successfully used for several organodichlorosilanes and organotrichlorosilanes or its mixtures [2,15-17] as well as for bis(trichlorosilanes) [18] or tetrachlorosilane [3,19] for producing highly cross-linked polysilylcarbodiimide xerogels. Gelation times, shrinkage, yield and pyrolysis behavior change with the kind of chlorosilane used as precursor (Table 2).

Table 2. Comparison of the gelation and aging behavior as well as the ceramic yields for different carbodiimide gels prepared in the presence of 0.1 equivalents of pyridine.

Starting chlorosilane	Idealized composition	Gelation time (d), 45°C	Shrinkage $\Delta L/L_o$ (%)	Ceramic yield (%)
$C_6H_5SiCl_3$	$[C_6H_5Si(NCN)_{1.5}]_n$	1.5	37	58
$Cl_3Si\text{-}SiCl_3$	$[Si_2(NCN)_3]_n$	0.5	38	54
$Cl_3Si(CH_2)_2SiCl_3$	$[Si(CH_2)_2Si(NCN)_3]_n$	1.5	37	62
$Cl_2MeSi\text{-}(CH_2)_2SiMeCl_2$	$[Me_2Si_2(CH_2)_2(NCN)_2]_n$	60	0	n.d.
$Cl_2MeSi\text{-}SiMeCl_2$	$[Me_2Si_2(NCN)_2]_n$	181	0	n.d.

n.d. = not determined.

References

[1] K. Komeya, M. Matsui. In Materials Science and Technology Vol. 11 (Eds: R. W. Cahn, P. Haasen, E. J. Kramer), Wiley VCH, Weinheim, 1994, p. 517.

[2] R. Riedel, E. Kroke, A. Greiner, A. O. Gabriel, L. Ruwisch, J. Nicolich, Chem. Mater. 1998, **10**, 2964.

[3] R. Riedel, A. Greiner, G. Miehe, W. Dressler, H. Fuess, J. Bill, F. Aldinger, Angew. Chem. Int. Ed. Engl. 1997, **36**, 603.

[4] W. Dressler, R. Riedel, Int. J. Refractory Metals Hard Mater. 1997, **15**, 13.

[5] A. Kienzle, A. Obermeyer, R. Riedel, F. Aldinger, A. Simon, Chem. Ber. 1993, **126**, 2569.

[6] A. Kienzle, Dissertation, Universität Stuttgart, Germany, 1994.

[7] D. Seyferth, C. Strohmann, N. R. Dando, A. J. Perrotta, J. P. Gardner, Mat. Res. Soc. Symp. Proc. 1994, **327**, 191.

[8] R. Riedel, A. Kienzle, W. Dressler, L. Ruwisch, J. Bill, F. Aldinger, Nature 1996, **382**, 796.

[9] H.-P. Baldus, M. Jansen, Angew. Chem. 1997, **109**, 338.

[10] A. O. Gabriel, Dissertation, Technische Universität Darmstadt, Germany, 1998.

[11] A. O. Gabriel, R. Riedel, S. Storck, W. F. Maier, Appl. Organomet. Chem. 1997, **11**, 833.

[12] A. S. Gordetsov, V. P. Kozyukov, I. A. Votokov, S. V. Sheludyakova, Y. I. Dergunov, V. F. Mironov, Uspekhi Khim., 1982, **51**, 848; Russian Chem. Rev. 1982, **51**, 485.

[13] J. Pump, U. Wannagat, Angew. Chem. 1962, **74**, 117.

[14] J. Pump, U. Wannagat, Liebigs Ann. Chem.1962, **652**, 21.

[15] E. Kroke, A. O. Gabriel, D. S. Kim, R. Riedel, in: From Molecules to Materials, Organosilicon Chemistry IV (Eds. N. Auner, J. Weis), Wiley-VCH, Weinheim, 2000, p. 812.

[16] A. O. Gabriel, R. Riedel, Angew. Chem. Int. Ed. Engl. 1999, **36**, 384.

[17] D. S. Kim, E. Kroke, R. Riedel, A. O. Gabriel, S. C. Shim, Appl. Organomet. Chem. 1999, **13**, 495.

[18] S. Nahar-Borchert, E. Kroke, R. Riedel, B. Boury, R. J. P. Corriu, J. Organomet. Chem. 2003, **686**, 127.

[19] Y.-L. Li, E. Kroke, A. Klonczynski, R. Riedel. Adv. Mater. 2000, **12**, 956.

Polyaniline – A Conducting Polymer

J. Stejskal and I. Sapurina

Abstract Polyaniline, a conducting polymer, is prepared by the oxidation of aniline hydrochloride with ammonium peroxodisulfate in aqueous medium. The polymer is obtained in nearly quantitative yield; its conductivity at 20°C is 4.4 Scm^{-1}. Protonated polyaniline is converted to non-conducting polyaniline base by treatment with ammonium hydroxide solution.

Classification

Form:	polymer powder
Function:	conducting polymer
Preparation:	oxidative polymerization
Composition:	$[C_{24}H_{18}N_4]_n \cdot 2n$ acid

Introduction

Conducting polymers have received ever-increasing attention, especially in last two decades.[1] Among them, polyaniline (PANI) is popular for its ease of preparation, good level of electrical conductivity, and environmental stability.[2–8] It is prepared by the oxidative polymerization of aniline with a suitable oxidant, *e.g.,* ammonium peroxodisulfate. The synthesis uses common chemicals and proceeds in acidic aqueous medium at ambient temperature and atmosphere. The polymer is produced within a few minutes as a precipitate and thus easily collected by filtration at virtually quantitative yield.

Polyaniline exists in a variety of forms that differ in chemical and physical properties.[1–8] The most common form, green protonated emeraldine (Fig. 1), has a conductivity on a semiconductor level in the order of 10^0 S cm^{-1}, many orders of magnitude higher than that of common polymers ($<10^{-9}$ S cm^{-1}) but lower than that of typical metals ($>10^4$ S cm^{-1}). Protonated PANI, so-called PANI "salt", *e.g.,* PANI hydrochloride, converts to the non-conducting blue emeraldine base when treated with a base [5,9] (Fig. 1), such as ammonium hydroxide.

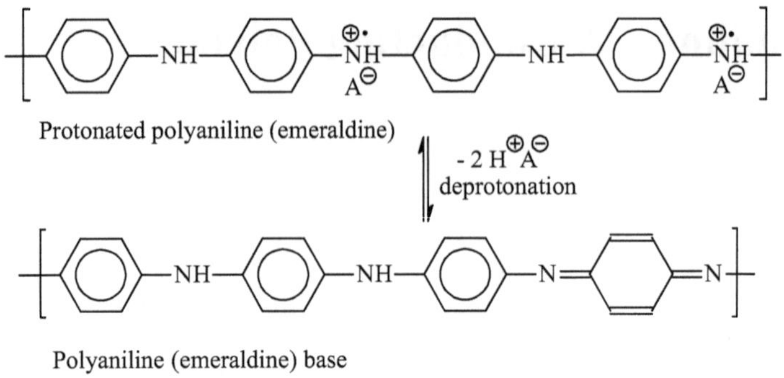

Fig. 1. Polyaniline (emeraldine) salt is deprotonated in alkaline medium to polyaniline (emeraldine) base. A^- is an anion, *e.g.*, chloride.

Polyaniline is used as a filler in the preparation of conducting composites, and for the surface modification of microparticles, powders, fibers, textiles, membranes, and porous substrates, endowing them with new electrical, chemical, and surface properties. The preparation of PANI colloids is one of the ways to cope with the difficult processibility of conducting polymers.[10,11] The changes in the physicochemical properties of PANI occurring in response to various external stimuli are used in various applications,[12] *e.g.*, in electrodes, sensors, and actuators. Some uses are based on the combination of electrical properties typical of semiconductors with materials parameters characteristic of polymers, like the development of "plastic" microelectronics and "smart" fabrics. Conducting polymers have been used in the design of new catalysts for organic syntheses. Surface coating with conducting polymers can modify adsorption phenomena and therefore be used in the separation science.

The preparation of PANI has recently been investigated within the collaborative project carried out by a task group of the International Union of Pure and Applied Chemistry.[13] The information provided here is based on the results and experience gained during that study. The follow-up project concerned the preparation of colloidal PANI dispersions and *in-situ* polymerized thin films. [14]

Materials

- Aniline hydrochloride, purum, used as received.
- Ammonium peroxodisulfate, purum, used as received.
- Acetone, used as received.
- Ammonium hydroxide solution (2–3 %).

Safety and Disposal

The oxidation of aniline is exothermic. Polymerization using aniline concentrations over 1 M, especially when carried out in large volumes (over 0.5 l), can result in the overheating of the system, followed by an explosion.[13] Such reaction conditions should be avoided.

Procedures

The polymerization of aniline reported here was designed to be as simple as possible. The synthesis is based on mixing aqueous solutions of aniline hydrochloride and ammonium peroxodisulfate at room temperature, followed by the separation of PANI hydrochloride precipitate by filtration and drying.

A. Preparation of Polyaniline Hydrochloride

The preparation of protonated PANI, here PANI hydrochloride, is based on the oxidation of 0.2 M aniline hydrochloride with 0.25 M ammonium peroxodisulfate in aqueous medium.[13] Aniline hydrochloride (2.59 g, 20 mmol) is dissolved in distilled water in a volumetric flask to 50 ml of solution. Ammonium peroxodisulfate (5.71 g, 25 mmol) is similarly dissolved in water also to 50 ml of solution. Both solutions are mixed at room temperature (~18–24°C) in a beaker, and left at rest or at gentle stirring to polymerize. After the polymerization has been completed in about 10 min, the mixture is left to cool down for several hours. The PANI precipitate is collected on a filter, washed with three 100 ml portions of 0.2M HCl, and similarly with acetone. Polyaniline (emeraldine) powder is dried in air and then *in vacuo*. The average yield of PANI hydrochloride is 2.13 g (98%).[13]

Mechanism of the Aniline Polymerization

The aniline hydrochloride, or generally any aniline salt (**1**), is oxidized at first to the aniline cation radical (**2**) (Fig. 2a). Although the detailed reaction mechanism of aniline polymerization is not fully understood, the formation of the protonated pernigraniline intermediate (**3**) (Fig. 2b) is observed during the polymerization and manifested by the deep blue color of the reaction mixture. At the end of polymerization, the pernigraniline is reduced with residual aniline to the final product, the green emeraldine form of PANI (**4**) (Fig. 2c). During the reaction, the ammonium peroxodisulfate is reduced to ammonium sulfate (Fig. 2d). Summing all reaction steps (**Σ**), the stoichiometric oxidant/monomer ratio 5/4 = 1.25 is found.[5] This is why the concentrations of aniline hydrochloride and ammonium peroxodisulfate were selected in the present protocol as 0.2 M and 0.25 M, respectively.

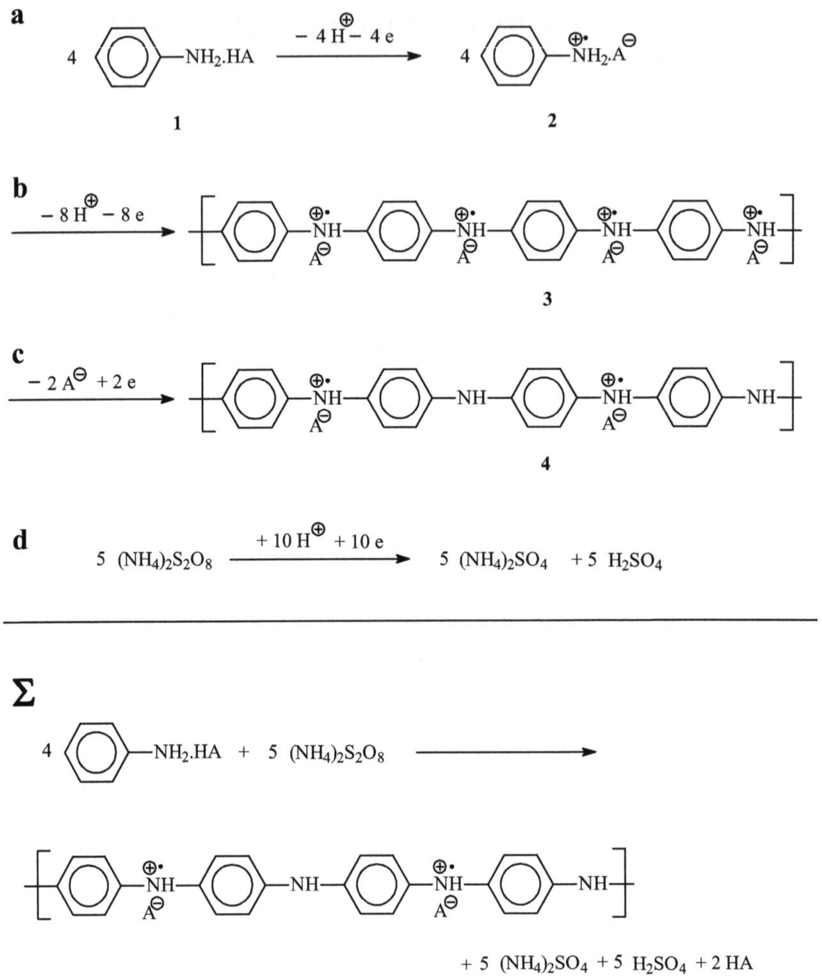

Fig. 2. Oxidation of an aniline salt by ammonium peroxodisulfate to yield protonated polyaniline (emeraldine) hydrochloride. HA is any acid.

The oxidation of aniline is exothermic and can conveniently be followed by the temperature changes [13,15] (Fig. 3). During the induction period, the temperature stays virtually constant, the reaction mixture becomes blue as oligomeric intermediates are produced. Once the polymerization has started, the temperature increases, the color of the mixture turns to deep blue, and the consistency becomes that of a slurry. The surface of the reaction vessels acquires a metallic tint due to a PANI coating. The course of polymerization can also be followed by changes in the acidity because protons (sulfuric acid) are produced during the polymerization (Fig. 2).

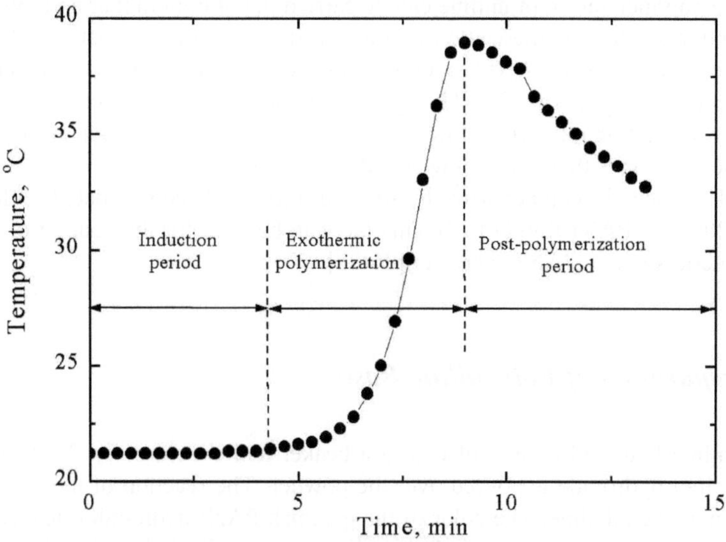

Fig. 3. Temperature profile in the polymerization of aniline (0.2 M aniline hydrochloride oxidized with 0.25 M ammonium peroxodisulfate in 100 ml of aqueous medium).

Comments

(1) The purity of the chemicals is not crucial as far as the yield and properties of PANI are concerned. The course of the polymerization is, however, accelerated by traces of various compounds.[16]

(2) An equimolar mixture of aniline and hydrochloric acid can be used instead of aniline hydrochloride. The presence of excess (1 M) hydrochloric acid in the reaction mixture improves the conductivity of PANI.[13]

(3) Various inorganic and organic acids at various concentrations can be used instead of hydrochloric acid in the polymerization of aniline.[17] The electrical and material properties of PANI vary correspondingly. Polyaniline is produced as fused nanogranules. [13] Polyaniline nanotubes are obtained when the oxidation of aniline takes place in the solution of weak acids, such as acetic acid [18,19] or in water. [20]

(4) When using ammonium peroxodisulfate as an oxidant, sulfuric acid is produced during the polymerization (Fig. 2). This means that the PANI is partly protonated also by this acid. Washing of PANI with hydrochloric acid after the preparation should replace most of the sulfate counter-ions with chloride and the resulting product is thus PANI hydrochloride. Subsequent rinsing with ace-

tone is needed to obtain PANI as a powder. Drying PANI precipitate while it still contains water produces polymer lumps, which may be difficult to process further.

(5) The polymerization of aniline can be carried out at both higher and lower temperatures. The polymerization is often carried out in an ice bath (0–2°C).[13] The thus produced PANI has a higher molecular weight, but its conductivity is improved only marginally.[13,21] The polymerization can be carried out in the frozen reaction mixture, below -10°C,[22,23] and proceeds even at -50°C. [21] The reaction is then much slower and takes several days.

(6) All surfaces in contact with the reaction mixture become coated with a thin (~200 nm) PANI film.[14,25] This fact can be used for the coating of various materials with a PANI overlayer.[26-28]

B. Preparation of Polyaniline Base

Polyaniline hydrochloride is placed in a beaker and excess of *ca* 2–3% aqueous ammonium hydroxide is poured over the powder. The reaction of the supernatant liquid must be alkaline. The color of the greenish PANI hydrochloride changes to blue after neutralization (Fig. 1). Polyaniline base is collected on the filter, washed with the solution of ammonium hydroxide, followed by acetone, and then dried.

Characterization

Elemental composition: The chlorine content reflects the protonation in PANI hydrochloride (Table 1), the presence of sulfur corresponds to a partial incorporation of residual sulfate or hydrogen sulfate anions produced by the reduction of peroxodisulfate during polymerization (Fig. 2).

Table 1. Elemental composition of polyaniline [13]

Sample	% C	% H	% N	% Cl	% S
Polyaniline hydrochloride					
Found	59.7	4.9	10.6	11.1	1.0
Calcd.[a]	66.2	4.6	12.9	16.3	-
Polyaniline base (after deprotonation of polyaniline hydrochloride)					
Found	75.0	5.0	13.9	0.6	0.3
Calcd.[a]	79.5	5.0	15.5	-	-

[a] Based on the formulae shown in Fig. 1.

Hydrochloric acid is removed from the PANI hydrochloride after deprotonation with ammonium hydroxide (Fig. 1). The relative proportion of carbon and nitrogen in the PANI base is thus increased, at the expense of the lower content of chlorine. Some chlorine remains in the PANI base even after deprotonation, indicating partial substitution of the phenyl rings with chlorine.[24] Sulfonation of the phenyl rings is responsible for the presence of sulfur in the PANI base.[20]

FTIR spectra: The infrared spectrum of PANI hydrochloride shows a broad absorption at wavenumbers >2000 cm^{-1}, which is characteristic of the conducting form of PANI [29,30]. Typical peaks in the infrared spectra of PANI hydrochloride, corresponding to quinone and phenyl ring deformations, are observed at 1569 cm^{-1} and 1480 cm^{-1} (Fig. 4). These are blue-shifted to 1590 cm^{-1} and 1500 cm^{-1} after deprotonation to PANI base.[31,32] The band at 1374 cm^{-1}, associated with C–N stretching in the neighborhood of a quinonoid ring, is present in the spectrum of PANI base but absent from the spectrum of PANI hydrochloride. The absorption at 1302 cm^{-1} corresponds to π-electron delocalization induced in the polymers by protonation [33] and is reduced after the deprotonation. The band characteristic of the conducting protonated form is found at about 1245 cm^{-1}. The band at 1144 cm^{-1} can be assigned to a vibration mode of a protonated imine group. It overlaps the band of in-plane C–H deformation vibrations at 1164 cm^{-1} observed in PANI base.[32,34-36] The aromatic-ring and out-of-plane C–H deformation vibrations manifest themselves in the region of 900–700 cm^{-1}.

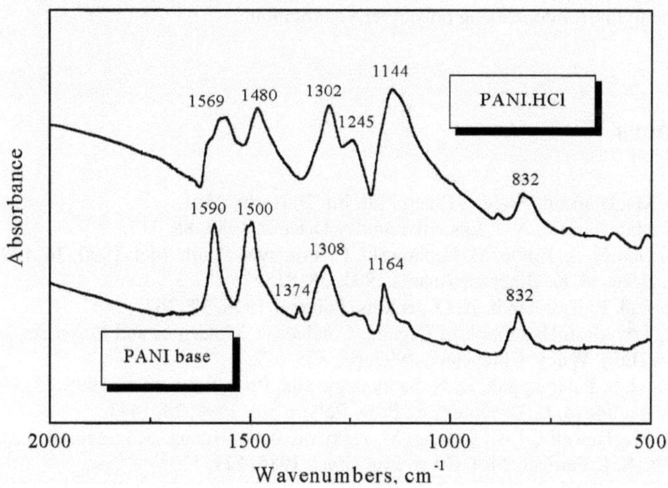

Fig. 4. Infra-red spectra of polyaniline hydrochloride and polyaniline base dispersed in potassium bromide pellets.

Molar mass: The mass-averaged molar mass of PANI base determined by gel permeation chromatography in *N*-methylpyrrolidone by using the polystyrene calibration is $M_w = 58\ 100$ g mol^{-1} (Fig. 5). This is the value corresponding to a degree of polymerization of about 640 aniline units, a value common for the many polymers met in practice. The molar mass distribution is relatively broad, the mass-to-number molar-mass ratio being $M_w/M_n = 3.3$.

Density: The average density of PANI hydrochloride is 1.329 ± 0.027 g cm^{-3} at 20 °C and that of PANI base 1.245 ± 0.006 g cm^{-3}.[13]

Conductivity: The average conductivity of PANI hydrochloride at 20 °C found on 59 independently prepared samples compressed into pellets [13] was 4.4 ± 1.7 S cm^{-1}. The conductivity of PANI base was many orders of magnitude lower, *viz.* $(6.0 \pm 1.8)\times10^{-11}$ S cm^{-1}.

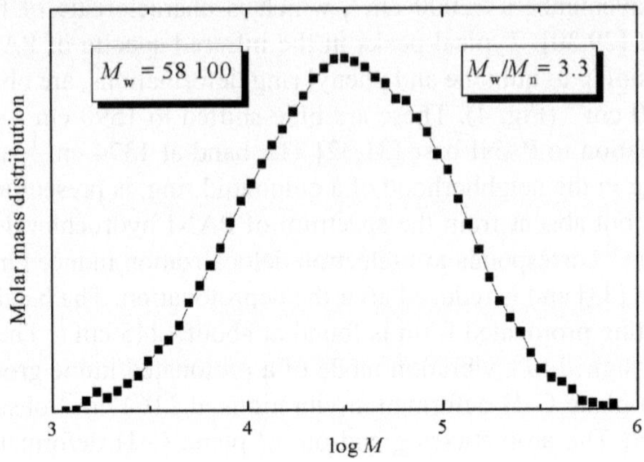

Fig. 5. Molar mass distribution of polyaniline base determined by gel permeation chromatography in *N*-methylpyrrolidone using polystyrene calibration.

References

[1] A. G. MacDiarmid, Angew. Chem., Int. Ed. 2001, **40**, 2581.
[2] A. G. MacDiarmid, A. J. Epstein, Faraday Discuss. 1989, **88**, 317.
[3] E. M. Geniès, A. Boyle, M. Lapkowski, C. Tsintavis, Synth. Met. 1990, **36**, 139.
[4] A. A. Syed, M. K. Dinesan, Talanta 1991, **38**, 815.
[5] J. Stejskal, P. Kratochvil, A. D. Jenkins, Polymer 1996, **37**, 367.
[6] D. C. Trivedi in Handbook of Organic Conductive Molecules and Polymers, Vol. 2, H. S. Nalwa (Ed.), Wiley, Chichester, 1997, pp. 505–572.
[7] J. Anand, S. Palaniappan, D. N. Sathyanarayana, Prog. Polym. Sci. 1998, **23**, 993.
[8] N. Gospodinova, L. Terlemezyan, Prog. Polym. Sci. 1998, **23**, 1443.
[9] A. G. MacDiarmid, J.-C. Chiang, M. Halpern, W.-S. Huang, S. L. Mu, N. L. D. Somasiri, W. Wu, S. I. Yaniger, Mol. Cryst. Liq. Cryst. 1985, **121**, 173.
[10] J. Stejskal, J. Polym. Mater., 2001, **18**, 225.
[11] J. Stejskal in Dendrimers, Assemblies, Nanocomposites, The MML Ser. Vol. 5, R. Arshady, A. Guyot (Eds.), Citus Books, London, 2002, pp. 195–281.
[12] Handbook of Conducting Polymers, T. A. Skotheim, R. L. Elsenbaumer, J. R. Reynolds (Eds.), Dekker, New York, 1998, pp. 823–1073.
[13] J. Stejskal, R. G. Gilbert, Pure Appl. Chem. 2002, **74**, 857.
[14] J. Stejskal, I. Sapurina, Pure Appl. Chem. 2005, **77**, 815.
[15] Y. Fu, R. L. Elsenbaumer, Chem. Mater. 1994, **6**, 671.
[16] J. Stejskal, P. Kratochvíl, M. Špírková, Polymer 1995, **36**, 4135.
[17] J. Stejskal, D. Hlavatá, P. Holler, M. Trchová, J. Prokeš, I. Sapurina, Polym. Int. 2004, **53**, 294.
[18] E. N. Konyushenko, J. Stejskal, I. Šeděnková, M. Trchová, I. Sapurina, M. Cieslar, J. Prokeš, Polym. Int. 2006, **55**, 31.

[19] J. Stejskal, I. Sapurina, M. Trchová, E. N. Konyushenko, P. Holler, Polymer 2006, **47**, 8253.
[20] M. Trchová, I. Šeděnková, E. N. Konyushenko, J. Stejskal, P. Holler, G. Ćirić-Marjanović, J. Phys. Chem. B 2006, **110**, 9461.
[21] J. Stejskal, A. Riede, D. Hlavatá, J. Prokeš, M. Helmstedt, P. Holler, Synth. Met. 1998, **96**, 55.
[22] L. H. C. Mattoso, A. G. MacDiarmid, A. J. Epstein, Synth. Met. 1994, **68**, 1.
[23] P. N. Adams, P. J. Laughlin, A. P. Monkman, A. M. Kenwright, Polymer 1996, **37**, 3411.
[24] G. M. Morales, M. Llusa, M. C. Miras, C. Barbero, Polymer 1997, **38**, 5247.
[25] J. Stejskal, I. Sapurina, J. Prokeš, J. Zemek, Synth. Met. 1999, **105**, 195.
[26] E. N. Kazantseva, J. Vilčáková, V. Křesálek, P. Sáha, I. Sapurina, J. Stejskal, J. Magn. Magn. Mater. 2004, **269**, 30.
[27] E. N. Konyushenko, J. Stejskal, M, Trchová, J. Hradil, J. Kovářová, J. Prokeš, M. Cieslar, J.-Y. Hwang, K.-H. Chen, I. Sapurina, Polymer 2006, **47**, 5715.
[28] N. V. Blinova, J. Stejskal, M. Trchová, G. Ćirić-Marjanović, I. Sapurina, J. Phys. Chem. B 2007, **111**, 2440.
[29] A. J. Epstein, J. M. Ginder, F. Zuo, R. W. Bigelow, H. S. Woo, D. B. Tanner, A. F. Richter, W. S. Huang, A. G. MacDiarmid, Synth. Met. 1986, **16**, 303.
[30] Z. Ping, J. Chem. Soc., Faraday Trans. 1996, **92**, 3063.
[31] M. Trchová, I. Šeděnková, E. Tobolková, J. Stejskal, Polym. Degrad. Stab. 2004, **86**, 179.
[32] I. Šeděnková, J. Prokeš, M. Trchová, J. Stejskal, Polym. Degrad. Stab. 2008, **90**, in press.
[33] J. C. Chiang, A. G. MacDiarmid, Synth. Met. 1986, **13**,193.
[34] X. R. Zeng, T. M. Ko, Polymer 1998, **39**, 1187.
[35] S. Quillard, G. Louarn, S. Lefrant, A. G. MacDiarmid, Phys. Rev. B 1994, **50**,12496.
[36] M. Cochet, G. Louarn, S. Quillard, M. I. Boyer, J. P. Buisson, S. Lefrant, J. Raman Spectrosc. 2001, **31**, 1029.

[19] J. Stejskal, I. Sapurina, M. Trchová, E. N. Konyushenko, P. Holler, Polymer 2006, 47, 8253.
[20] M. Trchová, I. Šeděnková, E. N. Konyushenko, J. Stejskal, P. Holler, G. Ćirić-Marjanović, J. Phys. Chem. B 2006, 110, 9461.
[21] J. Stejskal, A. Riede, D. Hlavatá, J. Prokeš, M. Helmstedt, P. Holler, Synth. Met. 1998, 96, [illegible]
[illegible]

Allyl- and Hydroxytelechelic Poly(isobutylenes)

W. H. Binder and R. Zirbs

Abstract Telechelic polyisobutylenes with low polydispersities and defined chain lengths can be obtained by quasiliving cationic polymerization of isobutylene in high yields up to molecular weights of 10^5 g mol^{-1}. Ally- and hydroxyl-endgroups can be introduced quantitatively.

Classification

form:	liquid polymer
function:	elastomer
preparation:	quasiliving cationic polymerization
composition:	$[C_4H_8]_n$

Introduction

Polyisobutylene (PIB) is one of the most important technical speciality polymers and can be obtained by cationic polymerization of isobutene.[1] PIB is characterized by unique properties – among them low glass transition temperature (T_g ~ -60°C), chemical resistance, good thermal and oxidation stability and biocompatibility. Its main use is focussed on butyl elastomers, where it constitutes an important part of butyl rubber as a copolymer with isoprene, yielding products with applications for tires, cable coatings, beltings and hoses.

The synthetic approach of poly(isobutylene) can be achieved exclusively by the cationic polymerization of isobutylene,[2] either in a nonliving or a living process. The living cationic polymerization of isobutylene [3,4] is one of the important achievements, the major results are compiled in the reviews given.[4, 5] The polymerization is of a quasiliving type initiated by tertiary chlorides,[6] tertiary alcohols,[7] tertiary esters,[8] tertiary ethers,[9] tertiary peroxides [10] and epoxides [11] (Scheme 1). The use of multivalent initiators controls the molecular architecture giving way to the preparation of block- and star copolymers. Thus multivalent cumyl chlorides, acetoxides, and ethers as well as *tert*-γ-lactones, *tert*-alklychlorides and *tert*-epoxides can be used as initiating systems.

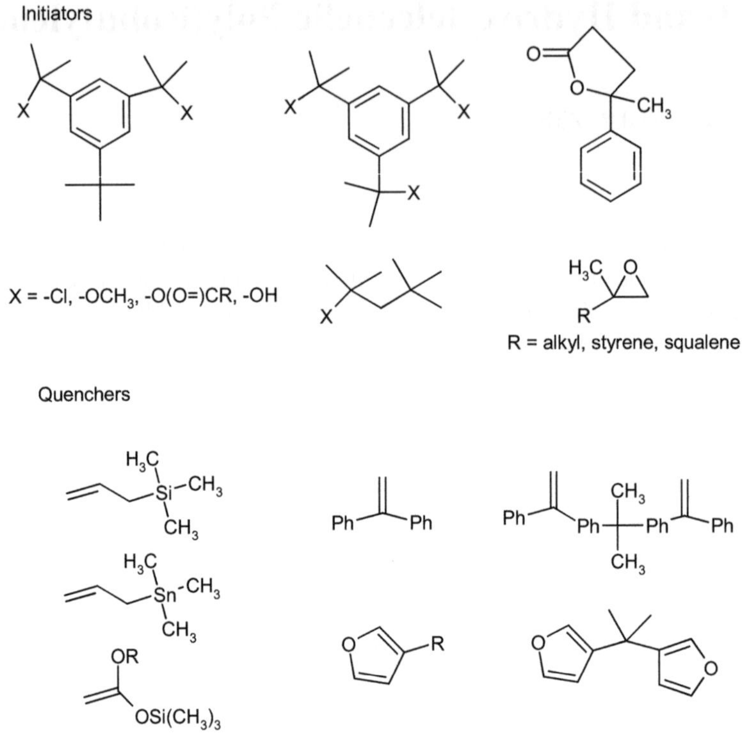

Scheme 1. Initiators and quenchers for the cationic polymerization of isobutylene

Usually Lewis acids [12] such as $AlCl_3$, Et_2AlCl, $TiCl_4$ or BCl_3 are used in non-polar, aprotic solvents (hexane, pentane, or mixtures with dichloromethane) together with a range of additives (DMA, DMSO, N-methylpyrrolidines)[13] and strong bases (sterically hindered pyridines)[14] serving to push the equilibrium towards the less reactive carbocationic species and additionally serve as proton traps during the polymerization reaction. Quenching of the living chain is achieved via allylsilanes,[15] and -stannanes;[16] silylketene acetals,[17] 1,1-diphenylethylenes[18] and substituted furanes.[19] The latter two quenchers can lead to the grafting of two and more polyisobutylene (PIB) chains onto the quencher molecule.[20]

Since within the living cationic polymerization of PIB polar residues cannot be introduced by direct methods, reaction with 1,1-diphenylethylene yields a stable cationic intermediate, which can be quenched subsequently with liquid ammonia,[21] alcohols[22] and acrylates [23] to yield the corresponding amino-, alkoxy- and acrylate telechelic PIB.

Usually PIB-OH and PIB-ene derivatives (Scheme 2) are starting points for the subsequent chemical transformations. Thus borane chemistry [24] in combination with nucleophilic substitutions;[25] Karstedt-type reaction [26] and ene-reaction [27] yield a variety of different synthetically useful PIB-derivatives. A complete overview about the functional groups achieved in polyisobutylenes is given by Kennedy in reference.[28] The present procedure gives an entry into allyl- and subsequently hydroxy-telechelic polyisobutylene in an easy reaction mode using a minimum of equipment. The products can be obtained with a polydispersity of 1.15 – 1.20.

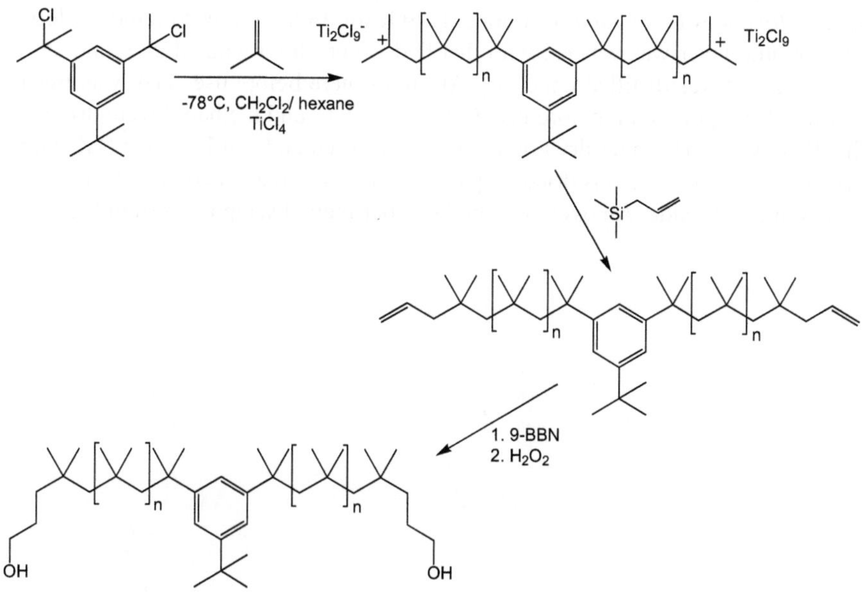

Scheme 2. PIB-OH and PIB-ene derivatives

Materials

- isobutylene (gas)
- 2,6-di-tert. butylpyridine
- N,N-dimethyl acetamide
- titanium (IV) chloride
- 1-tert-butyl-3,5-bis-(1-chloro-1-methylethyl)-benzene [29]
- trimethylallylsilane
- m-chloroperoxybenzoic acid
- 0.5M 9-BBN-solution in THF
- hydrogen peroxide
- tetrahydrofurane
- hexane
- methanol
- acetone
- dichloromethane

- distilled water

All materials were obtained from Aldrich and used without further purification if not mentioned otherwise. 1-tert-Butyl-3,5-bis-(1-chloro-1-methyl-ethyl)-benzene (DCCl) was obtained according to Faust *et al.*,[29] DMA (N,N-dimethylacetamide) was dried over calcium hydride and distilled in vacuo before use. n-Hexane was refluxed over conc. H_2SO_4 for 48 h in order to remove olefins. The organic layer was washed with distilled water, dried with $MgSO_4$ and stored over CaH_2. It was distilled under dry Ar-atmosphere before use. THF was freshly distilled from potassium before use. CH_2Cl_2, $CHCl_3$, and methanol were dried and distilled over CaH_2 under dry argon. DMF was dried and distilled over BaO under dry argon. Isobutylene was dried by passing the gas through a column packed with potassium hydroxide. An overview on the experimental setup is given in Figure 1.

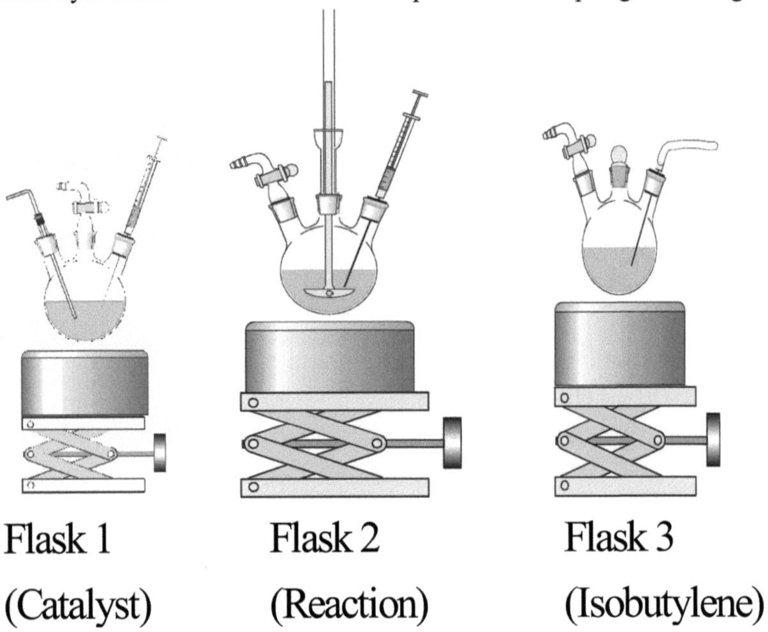

Fig. 1. Experimental setup

Safety and Disposal

Extreme care has to be taken with liquid isobutylene, since it is highly flammable. Additionally the concentrated sulfuric acid for the de-olefination of hexanes should be handled with caution.

Procedures

A. Syntheses of Allyl-telechelic PIB (1)

A general method for synthesis of allyl-terminated PIB´s is shown by the synthesis of allyl-terminated PIB with M_n =2500 $gmol^{-1}$ according to Ivan *et al.*[30, 31] and our procedures.[31,22] Dichloromethane (160 mL), olefin-free n-hexane (160 mL), DMA (1.96ml) and DCCl (3.05g, 10.61 mmol) were added to a 1L three-necked flask (flask 2) equipped with a septum, a mechanical stirrer and a nitrogen inlet and cooled to -80°C. Liquid isobutylene (14 g, 250 mmol; obtained by charging gaseous isobutylene through a dry, cooled flask under argon at -40°C) was charged to the reactor by a syringe (transfer from flask 3 to flask 2). After 5 min of stirring, a cold solution of $TiCl_4$ (40.25 g, 212 mmol; T = -40°C)) and 2,6-di-*tert*-butylpyridine (0.2 mL) in methylene chloride (80 mL) and olefin-free hexane (200 ml) was transferred to the reactor by a transfer needle (from flask 1 to flask 2). The temperature was held at -80°C during the whole polymerization procedure. After 10 min, a second addition of isobutylene (9.82 g, 175 mmol) followed. 20 min later, the polymerization was terminated by the addition of allyltrimethylsilane (6.7 g, 58.6 mmol). After 30 min the mixture was poured into a vigorously stirred saturated aqueous $NaHCO_3$ solution and filtered through a pad of *celite* (a layer of mineral-clay over a filtration paper and a glass-frit with porosity 3). The organic layer was separated, washed 5 times with distilled water and dried over $MgSO_4$. The solvent was removed by rotatory evaporator. Then, the polymer was redissolved in a small amount of n-hexane and precipitated 2 times into acetone in order to remove excess allyltrimethylsilane. Finally the colorless sticky polymer was dried in vacuum. Yield: 25.2g (94%).

Alternatively, DMA and di-*tert.*-butyl-pyridine can be exchanged by one equivalent of N,N,N',N'-tetramethylethylenediamine – this may lead to lower polydispersities in the final product according to Ivan et. al.[32]

Characterization

The main and most important characterization is via ^{1}H NMR-spectroscopy. Here, the main resonances of the main polymer of the endgroups (0.79; 1.54; 1.83; 2.01; 5.00; 5.83 ppm) as well as of the central core (7.17 ppm) can be detected and the amount can be determined via integration. This yields a factor of endgroup-functionalization, which reaches 100% in the described case. Additionally, ^{13}C NMR spectroscopy is a valuable tool. In the latter case (^{13}C NMR) the resonances of the endgroups can only be detected with polymers of a molecular weight below 10,000 $gmol^{-1}$.

^{1}H NMR (400 MHz, $CDCl_3$): δ(ppm) 0.79 (s, 12H), 0.83-1.50 (m, 433H), 1.54 (s, 4H),1.83 (s, 4H), 2.01 (d, 4H), 5.00 (t, 4H), 5.83 (m, 2H), 7.17 (s, 3H); ^{13}C

NMR (50 MHz, $CDCl_3$): δ(ppm) 28.70-31.60, 32.29, 34.75, 37.51-39.50, 50.29, 55.72, 58.00-60.00, 116.75, 120.06, 121.15, 136.08, 148.5, 148.93.

The polydispersity of the polymer (M_w/M_n) is between 1.1 and 1.2 as measured via SEC (calibration by narrow PIB-standards in THF).

Comments

(1) An argeon atmosphere is required for all polymerization steps. If this is not achieved, moisture may condense to the internal walls during polymerization and inhibit the controlled polymerization reaction.
(2) The transfer of liquid isobutylene by a syringe is tricky and should be done as fast as possible, Since the temperature of the syringe is usually higher than the boiling point of the liquid isobutylene.
(3) The filtration over *celite* can be substituted by silica gel. In this case, however, often the filtration is more difficult, since the filter-cake may block the frit, leading to a very slow filtration. Alternatively, the precipitate may be aged by keeping over night at room-temperature and subsequent filtration.
(4) In no case a temperature of the polymerization above -65°C may be overruled. Higher temperatures lead to broader molecular weights and lower yields.
(5) The polymerization reaction is finished after the addition of allyltrimethylsilane. After addition a depolymerization reaction is not possible and the polymer is stable at temperatures up to at least 50°C.
(6) Variation of the initiator/monomer ratio can yield polymers with different molecular weights. In the ideal case (100 % polymerization), the degree of polymerization reaches DP = [n(isobutylene)]/[n(DCCl)]

B. Syntheses of Hydroxyl-telechelic PIBs (2).

A general method for the synthesis of hydroxy-terminated PIB´s is shown by the synthesis of hydroxyl-terminated PIB with M_n = 2500 $gmol^{-1}$. Allyl-terminated PIB **1** (7.5 g, 3 mmol) was dissolved in THF (430 mL), freshly distilled over potassium. The solution was sparged with argon for 5 min. A 0.5M 9-BBN-solution in THF (75 ml, 37.5 mmol) was added dropwise under dry argon atmosphere at room temperature. After 5 h of stirring the mixture was cooled to 0°C and methanol (2.1 mL) and m-chloroperoxybenzoic acid (47 g, 0.19 mol) were added carefully. The reaction was allowed to react for 10-15 h, then hexane (100 mL) and distilled water (100 mL) was added. The aqueous phase was saturated with potassium carbonate. The organic layer was washed 5 times with 50% aqueous methanol, 5 times with distilled water, separated and dried with sodium sulfate. After filtration the solvent was evaporated and the product dried under vacuum at ambient temperature. Yield: 7.5g (100%).

Characterization

The main and most important characterization is via ^1H NMR-spectroscopy. Here, the main resonances of the main polymer as well as the endgroups (0.79; 1.83; 3.62 ppm) as well as the aromatic moieties of the central initiator at 7.17 ppm can be detected and the amount can be determined via integration. This yields a factor of endgroup functionalization, which reaches 100% in the described case. Additionally, ^{13}C NMR-spectroscopy is a valuable tool. Only with polymers of a molecular weight below 10^4 $gmol^{-1}$, the resonances of the endgroups can be detected.

^1H NMR (400 MHz, $CDCl_3$): δ(ppm) 0.79 (s, 12H), 0.83-1.67 (m, 424H), 1.83 (s, 4H), 3.62 (t, 4H J=6.9Hz), 7.17 (s, 3H); ^{13}C NMR (50 MHz, $CDCl_3$): δ(ppm) 27.75, 30.76-31.62, 32.28, 34.76, 37.81-38.93, 41.41, 55.57, 58.54-59.50, 63.96, 120.06, 121.15, 148.5, 148.93. The polydispersity of the polymer (M_w/M_n) is between 1.1 and 1.2 (SEC).

Comment

Frequently, an incomplete oxidation-reaction is observed. Mostly, the reason relates to reagents of poor quality (9-BBN is instable after storage over a long time; m-chloroperbenzoic acid decomposes over time).

References

[1] For general information in poly(isobutylene) see: Encyclopedia of Polymer Science, 2004, Wiley & Sons.
[2] Y. Kwon, R. Faust, Adv. Polym. Sci. 2004, **167**, 107.
[3] J. P. Kennedy, B. Ivan, Designed Polymers by Carbocationic Macromolecular Engineering: Theory and Practice, Carl Hanser Publ., Munich 1992, 96.
[4] J. P. Kennedy, Makromol. Chem. Macromol. Symp. 1992, **60**, 1.
[5] D. Held, B. Ivan, A. E. Müller, F. de Jong, T. Graafland, ACS Symp. Ser. 1997, **665**, 63. J. E. Puskas, G. Kaszas, Progr. Polym. Sci. 2000, **25**, 403.
[6] B. Keszler, G. Fenyvesi, J. P. Kennedy, J. Polym. Sci. A 2000, **38**, 706.
[7] A. V. Lubnin, J. P. Kennedy, Pure Appl. Chem., 1995, A32, 191. M. Gyor, H.-C. Wang, R. J. Faust, Pure Appl. Chem. 1992, **A29**, 639.
[8] R. Faust, J. P. J. Polym. Sci. A, 1987, **25**, 1847.
[9] S. Coca, K. Matyjaszewski, J. Polym. Sci. A 1997, **35**, 3595. G. Kaszas, J. Puskas, J. P. Kennedy, Polym. Bull. 1987, **18**, 123. B. Wang, M. K. Mishra, J. P. Kennedy, Polym. Bull. 1987, **17**, 213.
[10] M. K. Mishra, Macromol. Symp. 1996, **107**, 243.
[11] J. Song, J. Modis, J. E. Puskas, J. Polym. Sci. A 2002, **40**, 1005.
[12] M. Bahadur, T. D. Shaffer, J. R. Ashbaugh, Macromolecules 2000, **33**, 9548.
[13] M. Sawamoto, Progr. Polym. Sci. 1991, **16**, 111. B. Ivan, Makromol. Chem. Macromol. Symp. 1993, **75**, 181.
[14] R. F. Storey, K. R. Choate, Macromolecules 1997, **30**, 4799.
[15] L. V. Nielsen, R. R. Nielsen, B. Gao, J. Kops, B. Ivan, Polymer 1997, **38**, 2529.
[16] S. Hadijkyricacou, R. Faust, Polym. Mater. Sci. Eng. 1997, **76**, 300.

[17] Y. Kwon, R. Faust, C. X. Chen, E. L. Thomas, Macromolecules 2002, **35**, 3348.
[18] R. R. Quirk, T. Yoo, Y. Lee, J. Kim, B. Lee, Adv. Polym. Sci. 2000, **153**, 67. H. Schlaad, Y. Kwon, R. Faust, H. Mayr, Macromolecules 2000, **33**, 743.
[19] S. Hadjiyriacou, R. Faust, Macromolecules 1999, **32**, 6393.
[20] Y. C. Bae, R. Faust, R. Macromolecules 1998, **31**, 9379.
[21] S. Hadjikyriacou, R. Faust, Polym. Mater. Sci. Eng. 1997, **76**, 300.
[22] J. Feldthusen, B. Ivan, A. E. Müller, Macromolecules 1998, **31**, 578.
[23] S. Hadjikyriacou, Z. Fodor, R. J. Faust, Pure Appl. Chem. 1995, **A32**, 1137.
[24] P. Dreyfuss, J. P. Kennedy, Anal. Chem. 1975, **47**, 771. B. Ivan, J. P. Kennedy, V. S. C. Chang, J. Polym. Sci. A 1980, **18**, 3177. B. Koroskenyi, R. J. Faust, Pure Appl. Chem. 1999, **A36**, 1879. B. Koroskenyim, R. J. Faust, Pure Appl. Chem. 1999, **A36**, 471. K. M. Lee, C. D. Han, Macromolecules 2002, **35**, 760.
[25] B. Keszler, G. Fenyvesi, J. P. Kennedy, J. Polymer. Sci. A 2000, **38**, 706.
[26] M. A. Sherman, J. P. Kennedy, J. Polym. Sci. A 1998, **36**, 1891. P. Kurian, S. Zschoche, J. P. Kennedy, J. Polym. Sci. A 2000, **38**, 3200. J. Shim, S. Asthana, N. Omura, J. P. Kennedy, J. Polym. Sci. A 1998, **36**, 2997.
[27] E. Walch, R. J. Gaymans, Polymer 1994, **35**, 1774.
[28] M. K. Mishra, J. P. Kennedy, Desk. Ref. Funct. Polym. 1997, 57.
[29] M. Gyor, H. C. Wang, R. Faust, Pure Appl. Chem. 1992, **29**, 639
[30] B. Ivan, J. P. Kennedy, J. Polym. Sci. 1990, **28**, 89.
[31] W. H. Binder, M. J. Kunz, C. Kluger, G. Hayn, R. Saf, Macromolecules 2004, **37**, 1749.
[32] P. G. Groh, B. Ivan, M. Szesztay, F. deJong, T. Graafland, Polym. Prepr. 2000, **41**, 1379.

Symmetrically and Unsymmetrically Substituted Phthalocyanines

M. J. Ferreira Calvete and M. Hanack

Abstract A symmetrically and an unsymmetrically substituted phthalocyanine was synthesized by template reaction between the correspondent dinitriles and the metal salts. Considerations were made regarding the mechanistic process, as well as product distribution in the case of the unsymmetrically substituted phthalocyanine, which was synthesized by statistical condensation. These types of phthalocyanines are very important materials for nonlinear optic applications and for further functionalization to produce e.g. phthalocyanine polymers.

Classification

form:	amorphous solids, powder
function:	optical limiting material (symmetrical phthalocyanine), functionalizing material (unsymmetrical phthalocyanine)
preparation:	template reaction and/or statistical condensation
composition:	$(C_{144}H_{176}N_8O_8)InCl$ (symmetrical phthalocyanine), $(C_{84}H_{114}N_8O_7)Ni$ (unsymmetrical phthalocyanine)

Introduction

Phthalocyanines (Pc) are widely used as pigment dyes in textiles and polymers.[1] They exhibit remarkable qualities like lightfastness, brightness and stability towards environmental influences. Phthalocyanines consist of a planar macrocycle with an 18 π-electron system, which is responsible for its known stability. Since their first discovery, these macrocycles have been the target of intensive investigation,[1,2] particularly considering their properties as dyes.[3,4] In recent years, research has been retargeted for applications in materials science,[5-9] including, as example, as molecular semi-conductors,[10,11]as liquid crystals,[12] as Langmuir-Blodgett films,[13,14] in optical-data storage,[15] in cancer therapy,[16] in fuel cells,[17] in photoelectrochemical cells,[18] in photovoltaic cells[19] and for nonlinear optics.[20,21]

This extensive use of phthalocyanines is due to their remarkable structural flexibility.[22] The coordination number of the square-planar phthalocyanine is four, but many of the metals, having higher coordination numbers, can contain a variety of axial ligands.[2] Phthalocyanines do not occur in nature, but they are structurally related to porphyrins such as haemoglobin, vitamin B_{12} or chlorophyll (see Fig. 1).

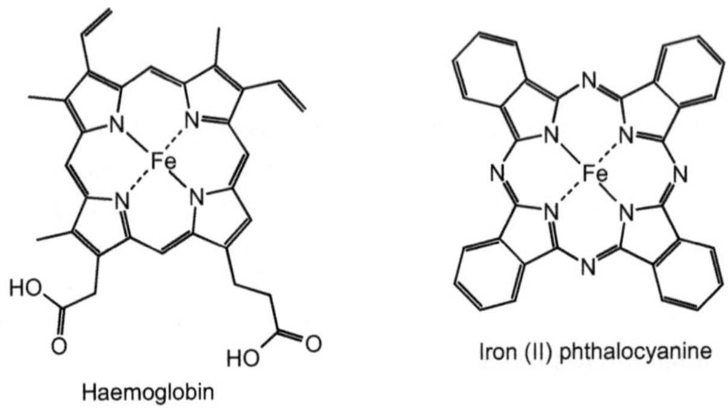

Fig. 1. Structural relationship between phthalocyanine and haemoglobin.

A metal-free phthalocyanine was found for the first time in 1907 as by-product during the preparation of 2-cyanobenzamide.[23] However, not much importance was given to the discovery at that time. Later, in 1927, a copper phthalocyanine was prepared in 23% yield by reacting 1,2-dibromobenzene with copper(I) cyanide in pyridine.[24] The structure of this substance was investigated meticulously by Linstead. He was the first to use the term phthalocyanine,[25] deriving the name from the Greek words *naphtha* (rock oil) and *cyanine* (blue). In the subsequent years he elucidated the structure of phthalocyanines as well as procedures for obtaining several metal Pc's and the metal free Pc's.[26-28]

In general, the synthesis of phthalocyanines proceeds from a single step reaction, by cyclotetramerization of benzoic acid or its derivatives, e.g. phthalic anhydride, phthalonitrile, phthalimide, *o*-cyanobenzamide, phthalonitriles or isoindolinediimine[2] in presence of metal salts (see Scheme 1).

Non substituted metal phthalocyanines are practical insoluble in common organic solvents. Solubility can be increased, for instance, by introduction of substituents in the periphery of the macrocycle (peripheral substitution). The most used precursor for the synthesis of substituted phthalocyanine is a substituted phthalonitrile, or in some cases, when the low reactivity of the precursor inhibits the macrocycle formation, substituted isoindolinediimines can be used as well.

The reaction mechanism for phthalocyanine formation is not yet fully understood. In any case it is generally assumed that the formation of the Pc's is controlled by a template effect (Scheme 2). Four phthalonitrile units coordinate in the first step to the metal ion.[29]

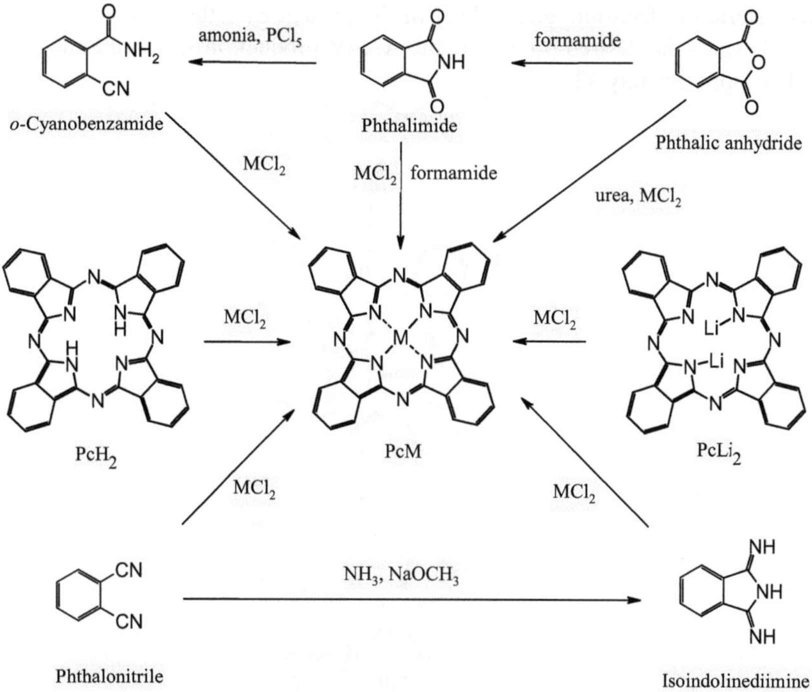

Scheme 1. General pathways for the preparation of phthalocyanine.

The combination of two different phthalonitriles permits the preparation of phthalocyanines with high functionality. In principle, two different phthalonitriles **A** and **B** can be condensed to give six different phthalocyanines, in a statistical distribution (Table 1).[30]

Table 1. Expected relative portions from the statistical condensation mixture of products (%).

A:B	AAAA	AAAB	ABAB	AABB	ABBB	BBBB
1:1	6.25	25	12.5	25	25	6.25
3:1	31.6	42.2	7.0	14.1	4.7	0.4
9:1	65.6	29.2	1.6	3.2	0.4	0.01
Permutations	1	4	2	4	4	1

By changing the ratio between two different phthalonitriles **A** and **B** in the statistical synthesis, the resulting amount of each isomer can be varied. When the ratio between the phthalonitriles **A** and **B** is 1:1 (i.e. 50%), the probability of obtaining **AA**, **AB** and **AB** is approximately $(0.5)^2 = 0.25$. However, for **AB** it must be considered that **BA** makes the same contribution, because there are two permutations of the elements **A** and **B** at two places. Thus, for all six specified Pcs, the probability is $(0.5)^4 = 0.0625$. This number, however, must be still multiplied by the number of permutations. This simple model does not consider the template and/or steric or electronic effects.[30] In the present case the stoichiometry used was 3:1, since the desired product is the **AAAB** product, in which the theoretical yield is approximately 42%.

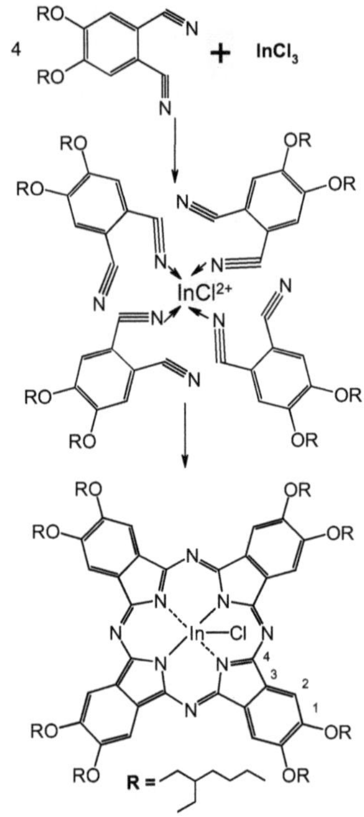

Scheme 2. Formation of the phthalocyanine macrocycle by metal induced coordination of the four phthalonitrile units (procedure A).

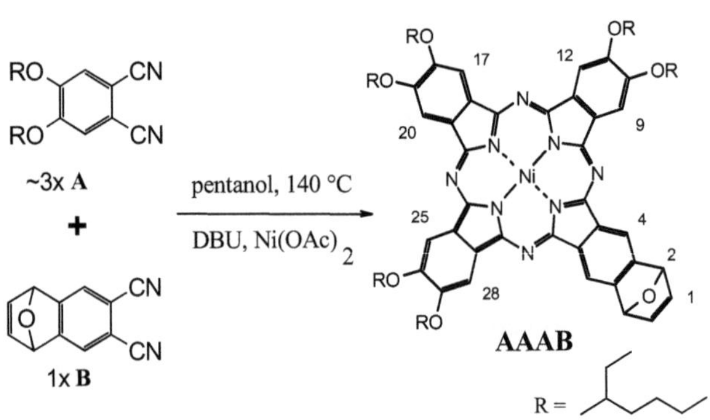

Plus **AAAA, AABB, ABAB, ABBB** and **BBBB** products

Scheme 3. Statistical synthesis of a functionalized **AAAB** phthalocyanine (procedure B). For sake of clarity only the **AAAB** product is shown. The mechanism for the formation of this material refers to Scheme 2.

Materials

- 4,5-Bis(2-ethylhexyloxy)-phthalonitrile was synthesized according to the literature procedure.[31,32]
- $InCl_3$, purity > 98%, purchased from Aldrich, used as received.
- 1-Chloronaphthalene, purity > 97%, purchased from Aldrich, used as received.
- 6,7-Dicyano-1,4-epoxy-1,4-dihydronaphthalene was synthesized according to the literature procedure.[31,32]
- 4,5-Bis(2-ethylhexyloxy)-phthalonitrile was synthesized according to the literature procedure.[31,32]
- $Ni(OAc)_2{\cdot}4H_2O$ purchased from Fluka, used as received.
- 1,8-Diazabicylo-[5,4,0]-undec-7-ene (DBU), purity > 98%, purchased from Aldrich, used as received.
- 1-Pentanol, purity > 95%, purchased from Aldrich, used as received.

All reactions were carried out under argon atmosphere. Additional purification procedures are described in the respective synthesis protocols. All solvents were purified and/or dried according to standard methods.

Safety and Disposal

Safety and handling instructions for the chemicals are found in the corresponding materials safety data sheets (MSDS). All manipulations should be carried out in a well-ventilated fume hood; protective gloves and safety glasses should be worn.

Procedure A (Scheme 2)

4,5-bis(2-ethylhexyloxy)-phthalonitrile (1.0 g, 2.6 mmol) and $InCl_3$ (150 mg, 0.7 mmol) was suspended in 3 mL of 1-chloronaphthalene and heated for 5 h at 185°C. After cooling down, the crude mixture was poured into 200 mL of methanol, stirred for 15 min and cooled in the refrigerator for a few hours. The precipitate was collected after centrifugation and washed with more methanol. The green solid was again dissolved in a small amount of CH_2Cl_2, methanol was added (~ 75 mL) and the CH_2Cl_2 evaporated. The solid was collected and washed with cold methanol to achieve further purification, followed by drying in vacuum at 90°C overnight.

[2,3,9,10,16,17,24,25-octa-(2-ethylhexyloxy)phthalocyaninato]-indium(III) chloride, green solid. Yield 560 mg, 50%.

Characterization

MS (FD): 1686.1 [M^+].
^1H NMR (THF-d_8): δ = 0.96 (s, br, 48 H, CH_3), 1.42, 1.63, (br, 64 H, CH_2), 1.97 (br, 8 H, CH), 4.36 (br, 16 H, OCH_2), 8.94 (br, 8 H, H-2). For NMR assignments see numbering in Scheme 2.
^{13}C NMR (THF-d_8): δ = 11.4, 14.1 (C-CH_3), 23.1, 24.1, 29.2, 30.8 (C-CH_2), 39.7 (C-CH), 72.2(C-OCH_2), 103.7, 105.4, 106.1 (br, C-2), 127.4, 129.8, 131 (br, C-3), 149.7, 150.3 152.8 (C-4), 154.2, 157.1(br, C-1).
UV/Vis (CH_2Cl_2): λ_{max} = 698.50, 671.5, 629.5, 401.5, 362.5 nm.

Comments

The structural prerequisite for the verification of NLO phenomena[33] in organic compounds, such as optical limiting, is the presence of a network of conjugated π–electrons, which infer high polarizability and fast charge redistribution when the conjugated molecule interacts with rapidly variable intense electromagnetic fields like those of laser radiations.[34]

Optical limiting is an important application of nonlinear optics, useful for the protection of human eyes, optical elements and optical sensors from intense laser pulses. An optical limiter is a device that strongly attenuates intense, potentially dangerous optical beams, while exhibiting high transmittance for low-intensity ambient light. In the variety of conjugated organic molecules possessing NLO properties, the class of phthalocyanines occupy a prominent position for the high thermal and chemical stability and the ease of preparation.[2]

The NLO properties of Pcs are of great interest, since these compounds can combine several physical and chemical properties which are favorable for the development of advanced NLO devices.

Varying properly the central atom (metal) in a phthalocyanine can introduce a change in the performance of the material as an optical limiter. Central moieties, such InCl, or InX, with X e.g. *para*-trifluoromethylphenyl, have the ability of introducing high dipole moments perpendicularly oriented to the Pc ring, which alter the electronic structure of the macrocycle, and new steric effects that modify the packing properties of PcMX's.[35]

Procedure B (Scheme 3)

An amount of 780 mg 6,7-dicyano-1,4-epoxy-1,4-dihydronaphthalene (4.0 mmol), 4.0 g of 4,5-bis(2-ethylhexyloxy)phthalonitrile (10.4 mmol) and 1.1 g of $Ni(OAc)_2 \cdot 4H_2O$ (4.43 mmol) were suspended in 30 mL pentanol, and a catalytic amount of DBU was added. The mixture was heated until 140°C and stirred for 20 h. After cooling, the mixture was poured in 150 mL methanol. The formed precipitate was isolated using centrifugation and washed several times with cold methanol. The crude mixture of the PcNi complexes was separated through chromatography on silica gel with CH_2Cl_2. After elution of fraction 1 [octa-(2-ethylhexyloxy)PcNi] (which was discarded), the **AAAB** product was obtained as the second fraction. Other subsequent fractions were also discarded. The solvent was removed, and the bluish-green solid was again dissolved in a small amount of CH_2Cl_2. Methanol was added (~75 mL) and the CH_2Cl_2 evaporated. The solid was collected and washed with cold methanol to achieve further purification, followed by drying in vacuum at 90°C overnight.
[2,3,9,10,16,17-hexa(2-ethylhexyloxy)-23,26-dihydro-23,26-epoxybenzophthalocyaninato]nickel, bluish-green solid. Yield: 900 mg, (18%).

Characterization

MS (FD): 1405.1 [M^+], 1389, 1293 [M^+-C_8H_{16}], 1180 [M^+-2 C_8H_{16}].
^1H NMR ($CDCl_3$): δ = 1.05, 1.18 (br, 36 H, CH_3), 1.52, 1.79 (br, 48 H, CH_2), 2.08 (br, 6 H, CH), 4.36 (br, 12 H, OCH_2), 6.24 (s, 2 H, H-2), 7.36 (s, 2 H, H-1),8.23 (3s, br, 6 H, H-9, H-12, H-17), 8.80 (s, 2 H, H-4). For NMR assignments see numbering in Scheme 3.
^{13}C NMR ($CDCl_3$): δ = 11.3, 11.5, 14.1, 14.3 (CH_3), 21.6, 23.2, 23.9, 24.2, 29.2, 29.4, 30.3, 30.7, 30.9 (CH_2), 39.7, 39.9 (CH), 71.8 (OCH_2), 82.7 (C-2), 103.8, 104.3, 104.5 (C-9, C-12, C-17), 113.4 (C-4), 130.5, 130.9 (C-8, C-13, C-16), 135.1 (C-5), 143.2, 143.6, 144.7, 146.0 (C-1, C-3, C-7, C-14, C-15), 149.3 (C-6), 151.9, 152.1, 152.5 (C-10, C-11, C-18).
UV/Vis (CH_2Cl_2): λ_{max} = 665, 601(shoulder), 309 nm.

Comments

Unsymmetrically substituted phthalocyanines with high solubility and with one or more functional groups are desirable building blocks for the preparation of e.g. semiconductive Pc-polymers,[36] linkage with other important materials, *e.g.* poly *p*-(phenylenevinylene) (PPV) and analogous polymers,[37,38] among other applications.[39,40] The unsymmetrical phthalocyanine here represented is the fundamental building block for any modulation in order to introduce the desired functional groups, usually applying a Diels-Alder strategy.[41]

References

[1] Phthalocyanines, Properties and Applications, (Eds: C. C. Leznoff, A. B. P. Lever), VCH, New York, 1989 – 1996, vol 1–4.
[2] M. Hanack, H. Heckmann, R. Polley, in Methoden der Organischen Chemie (Houben–Weyl), vol. E9d; 4th Ed.; Thieme Verlag, Stuttgart, 1997
[3] F. H. Moser, A. L. Thomas, The Phthalocyanines, CRC: Boca Raton, FL, 1983
[4] F. Baumann, B. Bienert, G. Rösch, H. Vollmann, W. Wolf, Angew. Chem. 1956, **68**, 133
[5] M. Hanack, M. Lang, Adv. Mater. 1994, **6**, 819
[6] M. Hanack, A. Datz, R. Fay, K. Fischer, U. Kepeler, J. Koch, J. Metz, M. Metzger, O. Schneider, H.-J. Schulze, in Handbook of Conducting Polymers, vol 1 (Ed.: T. A. Skotheim) M. Dekker Inc., New York, 1986, pp 133
[7] U. Drechsler, M. Hanack, in Comprehensive Supramolecular Chemistry, J. L. Atwood, J. E. D. Davies, D. D. McNicol, F. Vögtle Eds., Pergamon: Oxford, 1996, Vol 9, pp 283
[8] K.-Y. Law, Chem. Rev. 1993, **93**, 449.
[9] The Porphyrin Handbook, Aplications of phthalocyanines (Eds: K. M.; Kadish, K. M. Smith, R. Guilard,. Elsevier Science, San Diego, California, USA, 2003, vol. 19.
[10] J. Simon, J.-J. André, in Molecular Semiconductors, J. M. Lehn, C. W. Rees, Eds., Springer: Berlin, 1985, pp 73.
[11] T. J. Marks, Science 1985, **227**, 881.
[12] M. K. Engel, P. Bassoul, L. Bossio, H. Lehmann, M. Hanack, J. Simon, Liq. Cryst. 1993, **15**, 709.
[13] G. G. Roberts, M. C. Petty, S. Baker, M. T. Fowler, N. J. Thomas, Thin Solid Films 1985, **132**, 113.
[14] M. Burghard, M. Schmelzer, S. Roth, P. Haisch, M. Hanack, Langmuir 1994, **10**, 4265
[15] R. Ao, L. Kümmert, D. Haarer, Adv. Mater. 1995, **5**, 495.
[16] R. Bonnett, Chem. Soc. Rev. 1995, **95**, 19.
[17] A. B. P. Lever, M. R. Hempstead, C. C. Leznoff, W. Liu, M. Melnik, W. A. Nevin, P. Seymour, Pure Appl. Chem. 1986, **58**, 1467.
[18] D. Schlettwein, M. Kaneko, A. Yamada, D. Wöhrle, N. I. Jaeger, J. Phys. Chem. 1991, **95**, 1748.
[19] D. Wöhrle, D. Meissner, Adv. Mater. 1991, **3**, 129.
[20] M. Hanack D. Dini, M. Barthel, S.Vagin, Chem. Rec. 2002, **2(3)**, 129.
[21] M. J. F. Calvete, G. Y. Yang, M. Hanack , Synth. Met. 2004, **141**, 231. G. de la Torre, P. Vazquez, F. Agullo-Lopez, T. Torres, Chem. Rev. 2004, **104**, 3723.
[22] A. B. P. Lever, Adv. Inorg. Radiochem. 1965, **7** , 27
[23] A. Braun, J. Tscherniac, Ber. Dtsch. Chem. Ges. 1907, **40**, 270.
[24] H. de Diesbach, E. von der Weid, Helv. Chim. Acta 1927, **10**, 886.
[25] R. P. Linstead, Br. Ass. Adv. Sci. Rep. 1933, 465.
[26] R. P. Linstead, J. Chem. Soc. 1934, 1016.
[27] J. S. Anderson, E. F. Bradbrook, A. H. Cook, R. P. Linstead, J. Chem. Soc. 1938, 1151.
[28] R. P. Linstead, Ber. Dtsch. Chem. Ges. A 1939, **72**, 93.
[29] C. Rager, G. Schmid, M. Hanack Chem. Eur. J. 1999, **5**, 280.
[30] N. McKeown, I. Chambrier, M. Cook, J. Chem. Soc. Perkin Trans. 1990, **1**, 1169.
[31] M. J. F. Calvete, D. Dini, S. R. Flom, M. Hanack, R. G. S. Pong, J. S. Shirk, Eur. J. Org. Chem. 2005, **16**, 3499.
[32] R. Jung, M. Hanack, Synthesis 2001, **9**, 1386.
[33] Y.R. Shen, The Principles of Nonlinear Optics, J. Wiley & Sons, New York, 1984.
[34] B. Sheehy, L.F. Di Mauro, Ann. Rev. Phys. Chem. 1996, **47**, 463.
[35] J.S. Shirk, R.G.S. Pong, S.R. Flom, H. Heckmann, M. Hanack, J. Phys. Chem. A 2000, **104**, 1438.
[36] B. Hauschel, R. Jung, M. Hanack Eur. J. Inorg. Chem. 1999, **4**, 693.
[37] J. H. Burroughes, D. D. C. Bradley, A. R. Brown, R. N. Marks, K. Mackay, R. H. Friend, P. L. Burns, A. B. Holmes, Nature 1990, **347**, 539.

[38] S. C. Moratti, R. Cervini, A. B. Holmes, D. R. Baigent, R. H. Friend, N. C. Greenham, J. Gruener, P. J. Hamer, Synth. Met. 1995, **71**, 2117.
[39] M. J. F Calvete, D. Dini, M. Hanack, J. C. Sancho-Garcia, W. Chen, W. Ji,. J. Mol. Model. 2006, **12**, 543.
[40] M. Calvete, M. Hanack, Eur. J. Org. Chem. 2003, **11**, 2080.
[41] R. Jung, K.-H. Schweikart, M. Hanack, Eur. J. Org. Chem. 1999, **7**, 1687.

Index

MIX
Papier aus verantwortungsvollen Quellen
Paper from responsible sources
FSC® C105338

If you have any concerns about our products,
you can contact us on
ProductSafety@springernature.com

In case Publisher is established outside the EU,
the EU authorized representative is:
Springer Nature Customer Service Center GmbH
Europaplatz 3, 69115 Heidelberg, Germany

Printed by Libri Plureos GmbH
in Hamburg, Germany